Green Energy and Technology

For further volumes:
http://www.springer.com/series/8059

Wytze van der Gaast · Katherine Begg

Challenges and Solutions for Climate Change

Springer

Wytze van der Gaast
JI Network—JIN
Laan Corpus den Hoorn 300
Groningen, 9728 JT
The Netherlands

Dr. Katherine Begg
School of GeoSciences
University of Edinburgh
Kings Buildings, West Mains Rd
Edinburgh EH9 3JN
Scotland, UK

ISSN 1865-3529 ISSN 1865-3537 (electronic)
ISBN 978-1-4471-2602-7 ISBN 978-1-84996-399-2 (eBook)
DOI 10.1007/978-1-84996-399-2
Springer London Heidelberg New York Dordrecht

Printed on acid-free paper

Springer is part of Springer Science+Business Media (www.springer.com)

Acknowledgments

Around 5 years ago, we started a 2-year research project for the European Commission, as members of an international consortium. It focused on how international markets for emissions trading could generate extra (financial) support for clean technology transfer to developing countries. The project made it very clear that successful transfer of technologies is not just a matter of looking where the largest greenhouse gas emission reductions can be achieved, but requires looking at what a country needs for its sustainable development and how this can be done in a climate-friendly way. Equally important, it was clear that individual technology projects are not enough; system changes are needed so that technologies or measures can be rolled out at a scale required for making a difference.

We had an opportunity to further work on these insights through our involvement in the update of a *Handbook for Conducting Technology Needs Assessment for Climate Change*, under auspices of the UN Development Programme (UNDP). This handbook supports developing countries, through a step wise process, in working from a long-term vision on their sustainable development towards a long-term pathway to realise that vision with the lowest possible climate impacts. Working on this handbook, and our discussions with many experts, gave us a good opportunity to deepen our understanding of how achieving climate and development goals are interlinked with innovation processes in countries.

Through our work on preparing two background papers for the UNFCCC workshop on Technology Needs Assessments (Bonn, Germany, 1–2 June 2011), we were able to connect our insights on technology transfer with the ongoing international policy making for a new climate policy regime. How could an international climate regime enhance developing countries' capacity for technology innovation processes and could it facilitate the required funding?

In researching this book, the importance of the relationship between climate change, what we can do about it, and developing countries' efforts to achieve Millennium Development Goals, became clear. For instance, a changing climate makes achieving these goals increasingly difficult and expensive.

This book has drawn from these insights and we have presented these as five challenges. We would like to thank the European Commission (6th Framework

Programme), UNDP, and the secretariat of the UN Framework Convention on Climate Change for inviting us to become involved in the above projects. We would like to thank in particular Minoru Takada, Vladimir Hecl and Rona Wilkinson for their views, detailed discussions and pleasant co-operation over the years. We also thank Job Taminiau, Vlasis Oikonomou, Eise Spijker, John Henderson and Anna van der Gaast-Witkowska for contributing to and reviewing draft texts. Finally, our thanks go to the Springer team for their enthusiasm and support for this book.

Contents

Abbreviations

AAU	Assigned amount units
ACCO	Association of Climate Change Officers
AGECC	UN Secretary General's Advisory Group on Energy and Climate Change (AGECC)
AGF	UN High-level Advisory Group on Climate Financing
AWG KP	Ad hoc Working Group on continuation of the Kyoto Protocol
AWG LCA	Ad hoc Working Group on Long-term Cooperative Action
BASIC group	Group of countries with Brazil, South Africa, India and China
C	Celsius
CDM	Clean Development Mechanism of the Kyoto Protocol
CER	Certified emission reductions
CIC	Climate Innovation Centers
CIF	Climate Investment Fund
CO_2	Carbon Dioxide
CO_2-eq	Carbon Dioxide-equivalent
COP	Conference of the Parties to the UNFCCC
COP-MOP	Conference of the Parties serving as the Meeting of the Parties to the Kyoto Protocol
CSP	Concentrating solar power
CTCN	Climate Technology Centre and Networks
CTI	Climate Technology Initiative
DAC	OECD Development Assistance Committee
DfID	UK's Department for International Development
EGTT	Expert Group on Technology Transfer
EIT	European Institute of Innovation and Technology
ENTTRANS	Promoting Sustainable Energy Technology Transfers: Converting from a Theoretical Concept to Practical Action
ETS	Emissions trading scheme (e.g., EU ETS, US ETS)
EU	European Union
EUR	Euro (€)

FDI	Foreign Direct Investment
GEF	Global Environment Facility
GFCF	Gross Fixed Capital Formation
GHG	Greenhouse gases
IARU	International Association of Research Universities
IEA	International Energy Agency
INC	Intergovernmental Negotiating Committee for a Framework Convention on Climate Change
infoDev	World Bank Information for Development Program
IPCC	Intergovernmental Panel on Climate Change
JIQ	Joint Implementation Quarterly
KP	Kyoto Protocol
LCDS	Low carbon development strategies
LDC	Least Developed Countries
LDCF	Least Developed Country Fund
LEDS	Low emissions development strategies
MAC	Marginal abatement cost
MCDA	Multi criteria decision analysis
MDB	Multilateral Development Bank
MDGs	Millennium Development Goals
MRV	Measuring, reporting and verification
Mt	Megatonne
NAMAs	Nationally appropriate mitigation actions
NAPs	National adaptation plans
NAPAs	National adaptation programmes of action
NGO	Non-governmental organisation
NSI	National Systems of Innovation
NIMBY	Not in my backyard
ODA	Official Development Aid
OECD	Organisation for Economic Co-operation and Development
PFAN	CTI Private Financing Advisory Network
PISCES	Policy Innovation Systems for Clean Energy Security
RD&D	Research, development and demonstration
REDD	Reducing emissions from deforestation and forest degradation
REEEP	Renewable Energy and Energy Efficiency Partnership
Rio+20	United Nations Conference on Sustainable Development (or UNCSD)
SCCF	Special Climate Change Fund
SI	Systems Innovation
SME	Small and medium enterprises
TAP	Technology Action Plan
TEC	Technology Executive Committee
TNA	Technology Needs Assessment

UN	United Nations
UNCED	United Nations Conference on Environment and Development or Earth Summit
UNCSD	United Nations Conference on Sustainable Development (or Rio+20)
UNDP	United Nations Development Programme
UNEP	United Nations Environment Programme
UNFCCC	United Nations Framework Convention on Climate Change
US EPA	US Environmental Protection Agency
US$	US dollar
WMO	World Meteorological Organisation

Chapter 1
Challenge 1: Placing Climate Actions in a Wider Sustainable Development Context

Abstract Recent studies have shown an increasing urgency to address climate change and its impacts on ecosystems with the scale of action requiring no less than 'a green revolution.' A review of the climate negotiations and recent developments within the Cancun Agreements identifies five main challenges to progress. In this first chapter, the need for increased engagement of developing countries in global climate coalitions for action is discussed. It has become clear that climate policy making and sustainable development policies are increasingly becoming interlinked, especially in developing countries, where climatic changes affect achieving development goals and where climate action could improve energy access with low greenhouse gas emitting technologies for poverty alleviation. The challenge is to place the climate negotiations in this wider context of sustainability, equity and social change.

1.1 Why This Book?

Climate change and its accompanying consequences because of anthropogenic emissions of greenhouse gases (GHG) are among the most daunting environmental problems in the world today. It has become increasingly clear that higher GHG concentrations in the atmosphere through unchecked emissions provoke severe climate change and ocean acidification. Both impacts can fundamentally alter environmental structures on which humanity relies and have serious consequences for the food chain among others. Climate change, therefore, poses major socio-economic, technical and environmental challenges.

In 1992, climate change was an important agenda item at the UN Earth Summit, which resulted in an agreement on the UN Framework Convention on Climate Change (UNFCCC). In 2012, the UN summit 'Rio+20' will celebrate, among

W. van der Gaast and K. Begg, *Challenges and Solutions for Climate Change*, Green Energy and Technology, DOI: 10.1007/978-1-84996-399-2_1,

others, the 20th anniversary of the UNFCCC. In this time, much has been achieved, including a protocol (Kyoto) and several national and regional climate policy packages. However, recently, climate negotiations have been slowing down as countries seem reluctant to commit to ambitious next steps.

An important reason why the climate change focus has moved elsewhere has been the economic recession and the debt crisis. However, according to Jackson (2009), this has also diverted attention away from the key economic issue which is that the economic system we ascribe to has inherent weaknesses, such as its incompatibility with sustainability. He concludes that a wider more sustainable view has to be taken when looking for solutions.

Defence strategists have raised the alarm about the effects of climate change, energy and food shortages and socio economic problems, not only on the economy but also on international security (Rogers 2010). These key issues are interlinked. For instance, high CO_2 levels in the atmosphere can cause ocean acidification with effects on the marine ecosystem such as oxygen production and the food chain. Furthermore, climate change can have implications for many regions of the world. An example is effects on water resources and land use with changing rainfall patterns and temperatures. Sea level rise can affect the availability of land, and salt water ingress can affect existing water supplies. Therefore, Rogers (2010) argues that defence strategic thinking has to go further than conventional defence and look at the underlying causes, which are an economically polarised world and an environmentally constrained system.

Recent policy documents, such as the EU climate policy roadmap for 2050 (European Commission 2011) and the World Energy Outlook 2011 (IEA 2011), underline that the urgency of action on climate change has not diminished in any way. They suggest a way forward in which a mass change of attitudes and behaviour will be required, as well as technological change to low-emission sources and access to energy for all.

In developing countries, the interdependency between climate change and domestic policies is even greater as a changing climate directly affects people's well-being, such as limited access of poor people to energy (Practical Action 2010). With this in mind, there is every reason to address access to energy to reach the Millennium Development Goals, a priority goal for many developing countries, as a part of adopting low-emission technologies for avoiding climate change.

In other words, future agreements on climate change should be founded in equity and sustainability to maximise development and climate benefits from any investments made under the UNFCCC. We see this as a first challenge:

Challenge 1 To place the climate negotiations in the wider context of sustainability, equity and social change to develop agreements based on a comprehensive understanding that maximising the benefits in these areas can and should be achieved at the same time as decreasing GHG emissions.

Placing climate actions in a wider sustainable development context requires that countries develop a vision for their economic, social and environmental sustainable development goals and formulate a pathway towards that vision with low-emission and climate-resilient technologies and measures. The transfer of these technologies and measures is essential and has to be on a large scale, given the urgency of the problem. This leads us to two further challenges:

Challenge 2 To meet both countries' sustainable development and technology transfer needs by selecting technologies or measures for climate change mitigation and adaptation based on countries' sustainable development and climate goals.
Challenge 3 To create the low GHG emission and climate-resilient strategies and action plans to accelerate innovation needed to achieve sustainable development and climate goals on the scale and timescale required.

In Chaps. 2 and 3, we discuss how the second and third challenges could be fulfilled with the help of provisions developed under the UNFCCC for future climate policy making, such as assessment of technology needs, formulation of low-emission development strategies and development of national action plans for climate change mitigation and adaptation.

To support developing countries in formulating and implementing climate mitigation and adaptation actions in the context of sustainable development, recent climate negotiations have resulted in provisions and mechanisms for technology transfer, finance and capacity building. These are, however, far from being crystallised, which results in our fourth challenge:

Challenge 4 To rationalise the current directions in international climate policy making to provide coherent and efficient support to developing countries in devising and implementing strategies and action plans for low-emission technology transfers to deliver climate and sustainable development goals.

Finally, a key prerequisite for successful climate and sustainable development pathways is availability of funding. Not only will finance needs be large, given the massive tasks that the world faces, but they will also be diverse in form and source to address country-specific needs and needs for purchase of equipment or improving markets or systems. This leads us to the final challenge of this book:

Challenge 5 To facilitate development of an international framework for financial resources to support technology development and transfer, improve enabling environments for innovation, address equity issues such as poor people's energy access, and make implementation of activities possible at the desired scale within the country.

This chapter sets the stage for discussing these challenges in this book by briefly describing the scientific and political history of the climate change issue, how it has become increasingly interlinked with development issues and whether and how the new provisions and mechanisms in climate negotiations offer a way forward for countries' low-emission and climate-resilient development.

1.2 Climate Change as a Global Issue

1.2.1 (Re)gaining the Urgency

As climate change research progresses, there is an increasing understanding of the potential severity and seriousness of possible climate change impacts on the Earth's ecosystems. This has led the Intergovernmental Panel on Climate Change (IPCC) to conclude in its 2007 Fourth Assessment Report that average global temperatures should not rise by more than 2°C above pre-industrial times levels (IPCC 2007). This threshold of 2°C has been adopted in the official UN climate negotiations (i.e. the Cancun Agreements of December 2010, UNFCCC 2010, para 4) as the maximum allowable temperature increase to avoid irreversible damage to ecosystems around the world and the global climate.[1]

According to IPCC (2007), the atmospheric concentrations of the main GHGs now far exceed their natural range during the last 650,000 years, which is primarily because of anthropogenic use of fossil fuels. For instance, global GHG emissions have increased by 70% between 1970 and 2004 (IPCC 2007). Based on this trend, IPCC (2007) estimates a global average temperature increase during this century of between 1.1 (the low end of estimates) and 6.4°C (the high end) above pre-industrial times temperature levels.

The 2007 IPCC assessment report was published at a time of increasing awareness among people of the climate change issue. People started to realise potential impacts of climate change and the willingness to contribute to a solution even became a marketing instrument in several industrialised countries. For instance, meetings and plane tickets could be made "carbon neutral" by offsetting their contribution to GHG emissions through emission reduction projects elsewhere. An important stimulus for this increased awareness was undoubtedly given by the publication of "An Inconvenient Truth" by former US Vice-President Al Gore (Gore 2006). This publication and documentary placed recent GHG emission

---

[1] However, the Cancun Agreements acknowledge that the target of 2°C may not be sufficient and is notional. Recent calculations indicate that "if global GHG emissions are halved by 2050 relative to 1990, there is still a 12–45% estimated probability of exceeding 2°C—assuming a range of published climate sensitivity distributions. Emission levels in 2020 are a less robust indicator, but for the scenarios considered, the probability of exceeding 2°C rises to 53–87% if global GHG emissions are still more than 25% above 2000 levels in 2020" (Meinhausen et al. 2009).

trends in a long-term historical perspective, but, most importantly, it visualised potential consequences of climate change so that a thus far quite abstract discussion could suddenly be easily understood by large groups.

At this time, Gore and the IPCC received the 2007 Nobel Peace Prize. However, this momentum could not be maintained in the years to follow. During 2008, the financial crisis that followed the credit crunch seemed to bring people's attention back to everyday issues such as jobs, income and further uncertain financial prospects. Furthermore, reported mistakes in IPCC (2007) and false accusations that climate scientists intentionally overstated climate change evidence (Berkeley Earth Surface Temperature Project (2011) and Muir-Russell (2010)) caused attention to be turned away from the climate change debate.

Box 1.1 GHG emission trends 1990–2010
According to the Global Carbon Project,[2] annual growth of global and regional CO_2 emissions has accelerated from a rate of 1.1% during 1990–1999 to 3% per year for the 2000–2004 period. During 1990–2009, GHG emissions[3] in Australia rose by 30.4%, in Canada by 17.0% and in the USA by 7.2% UNFCCC-SBI (2011). The European Union managed to decrease GHG emissions by 17.4% during the same period, but this was largely caused by emission reductions in the Central and Eastern European Member States after the disintegration of their centrally planned economies in the early 1990s. Höhne et al. (2011) presented GHG emission development figures for the energy sector, including the impacts of the recession following the international financial crisis. They concluded that between 2005 and 2010, global energy-related emissions increased by 10%, from 30 to 33 Gigatonnes CO_2. It was found that while emissions in industrialised countries have reduced because of the recession, in developing countries there has been an increase, although the recession may have slowed down the increase. For instance, China's energy sector CO_2 emission grew during 2005–2010 by more than 40%; in India, the increase was 45% (Höhne et al. 2011).

However, the urgency of the increasing risk of severe, abrupt or irreversible climate change has not lessened, but increased as GHG emissions have been continuously rising (see Box 1.1). In fact, the latest scientific knowledge on climate change indicates that the world is on a GHG emissions trajectory that is worse than the IPCC (2007) worst case scenario and that there is a risk of severe disruption of the climate system. For instance, the International Association of Research Universities (IARU) congress Climate Change: Global Risks, Challenges and Decisions, held in Copenhagen in March 2009, concluded that:

---

[2] http://www.globalcarbonproject.org/misc/carbontrends.htm. Accessed 28 May 2011.

[3] These figures do not contain GHG emissions in the areas of land use, land-use change and forestry.

> Recent observations confirm that, given high rates of observed emissions, the worst-case IPCC scenario trajectories (or even worse) are being realized. For many key parameters, the climate system is already moving beyond the patterns of natural variability within which our society and economy have developed and thrived. These parameters include global mean surface temperature, sea-level rise, ocean and ice sheet dynamics, ocean acidification, and extreme climatic events. There is a significant risk that many of the trends will accelerate, leading to an increasing risk of abrupt or irreversible climatic shifts. (IARU 2009)

In addition, in a special issue of the *Royal Society journal* (New et al. 2011), it was shown that a continuation of GHG emissions growth according to business-as-usual trends could lead to a 4°C rise in global temperature by the 2060s. Since there was a 5°C decrease in global temperature at the last ice age, a positive change in temperature of that magnitude, occurring so quickly, implies serious disruption to all the systems on which our existence depends.

1.2.2 Climate Negotiations and Sustainable Development

The increasing insights on the socio-economic, technical and environmental risks of climate change have resulted in a range of international climate policy agreements. The first of these agreements was reached at the UN Earth Summit of 1992[4] with the adoption of the UN Framework Convention on Climate Change or UNFCCC, which established a framework for further international co-operation on climate policy making. Since then, much has been achieved in terms of climate policy making, such as the Kyoto Protocol of 1997, the Marrakech Accords of 2001 and the Cancun Agreements of 2010. These texts have offered guidance on ways to mitigate climate change and how to adapt to climatic changes.

However, climate change was not the only issue on the agenda of the 1992 Earth Summit. The overall focus was on the relationship between economic development and environmental degradation, such as climate change, threats to biodiversity and forest degradation. To address these concerns, other adopted documents, in addition to the UNFCCC, were:

- Agenda 21 as a programme for global action on sustainable development;
- The Rio Declaration on Environment and Development;
- The Statement of Forest Principles; and
- The Convention on Biological Diversity.

Together, these documents formed the message "that nothing less than a transformation of our attitudes and behaviour would bring about the necessary changes" (UN 1992).

---

[4] The official name of the meeting was: United Nations Conference on Environment and Development (UNCED). It was held in Rio de Janeiro, Brazil, on 3–14 June 1992. http://www.un.org/geninfo/bp/enviro.html. Accessed 27 August 2011.

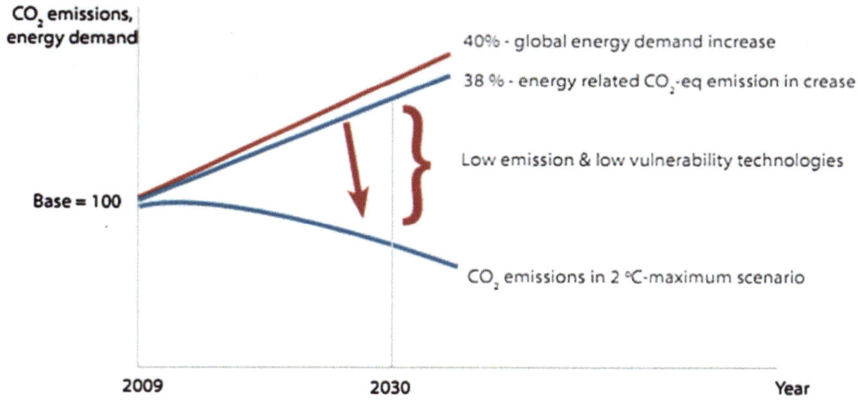

Fig. 1.1 The challenge the world faces to address climate change (Based on IEA 2009 and reproduced from UNDP 2010)

To mark the 20th anniversary of the 1992 Earth Summit, the UN Conference on Sustainable Development (UNCSD) or "Rio+20" (Rio de Janeiro, Brazil, 4–6 June 2012) will discuss progress made on these issues and explore gaps between goals and achievements. This should lead to a new policy agenda for a "green economy in the context of sustainable development and poverty eradication" and "creating an institutional framework for sustainable development."[5]

1.2.3 Inter-Dependencies Between Climate Change and Energy and Development Needs

The increasing insights on socio-economic, technical and environmental challenges posed by a changing climate have increasingly made clear that climate change cannot be considered in isolation from other issues. Instead, climate policy making has increasingly become interrelated with economic, financial, environmental and sustainable development policies, both in industrialised and developing countries. For example, efforts to reduce GHG emissions have to take place against the backdrop of an increasing global energy demand. According to the IEA (2009), global energy demand will increase by 40% around 2030. With the present status of energy technologies used and policies formulated, this will lead to a corresponding increase in GHG emissions (from 29 gigatonnes CO_2-equivalent to 40 gigatonnes or 38% increase). However, to limit average global temperature increases to 2°C, as explained earlier, global GHG emissions need to be roughly halved by 2050.

[5] http://www.uncsd2012.org/rio20/index.php?menu=17

Therefore, a strong breach with business-as-usual pathways is required and this is illustrated in Fig. 1.1.

Another example of the increasing interlinkage of climate change with other policy areas is that in developing countries, the urgency of climate change exists alongside the development urgencies as formulated by the Millennium Development Goals (MDGs). According to Practical Action (2010), one and a half billion people in the world have no access to electricity and three billion people rely on traditional biomass and coal for cooking. The impact of the latter on people's health is illustrated by the fact that indoor smoke from traditional cook stoves causes 1.4 million deaths per year. In general, Practical Action (2010) concludes that:

> Energy poverty is critically undermining the achievement of the [Millennium Development Goals, MDG]. As long as hundreds of millions of people remain deprived of the basic energy services needed to stay fed and healthy, earn a living, and allow the time needed for learning and fulfilment, the MDGs will remain out of reach.

In 2010, the UN Secretary General's Advisory Group on Energy and Climate Change (AGECC 2010) recommended universal access to modern energy services by 2030 (basic minimum threshold of modern energy services for both consumption and productive uses) and to reduce global energy intensity by 40% by 2030, by building and strengthening countries' capacity to implement effective policies, market-based mechanisms, business models, investment tools and regulations with regard to energy use. The Least Developed Countries Expert Group under the UNFCCC (LDC Expert Group 2009) has pointed out that development and adaptation policies and practice need to be blended.

These inter-dependencies show a need for strategies in countries whereby low GHG emission actions are identified in the light of countries' increasing energy demand and sustainable development priorities, including countries' need for reducing energy poverty.[6] This is reflected in the 2010 Cancun Agreements that encourage countries to formulate low-emission development strategies or plans in the context of their sustainable development (UNFCCC 2010, para 6), whereby actions to reduce GHG emissions should be "nationally appropriate" (UNFCCC 2010, Chap. III). Also, as explained earlier, the Rio+20 summit will focus on a green economy (i.e. including low GHG emission pathways) in the context of sustainable development.[7]

Before moving towards a discussion on how the individual climate and development perspectives of developed and developing countries could be addressed in an international climate policy framework, the next section first presents a brief synopsis of the history of climate negotiations and their dynamics, including the recent focus on aligning mitigation and adaptation measures with sustainable development goals.

---

[6] Note that in Practical Action (2010), energy access is considered for a range of services: lighting, cooking and water heating, space heating, cooling, information and communications and earning a living.

[7] http://www.uncsd2012.org/rio20/index.php?menu=17

1.3 Short History of Climate Negotiations

1.3.1 Increasing Climate Change Awareness During the 1980s

The adoption of the UNFCCC at the 1992 Earth Summit was not the starting point of international climate talks; rather, it was the conclusion of a series of governmental meetings and conferences that started already at the end of the 1970s. In 1979, the World Meteorological Organisation (WMO) organised, in Geneva (Switzerland), the First World Climate Conference that was primarily a scientific meeting to assess existing knowledge of how higher atmospheric GHG concentrations could lead to average temperature increases (WMO 1979). At the conference, model scenarios were presented based on meteorological data, data of GHG emissions and carbon sequestration in soils, forests and oceans, and corresponding development of atmospheric GHG concentration figures.

From a political perspective, human-induced global warming was in those days mainly considered a theoretical possibility, insufficiently backed by scientific evidence and surrounded by relatively large uncertainties (Arts 1998, p. 102). During the 1980s, climate models were further improved and awareness of the global warming issue among the public and politicians increased (Arts 1998, p. 103). In 1985, at the Villach Conference (9–15 October 1985, Villach, Austria), scientists called for the world to take action on climate change. This raised the profile of the whole issue and engaged a wider audience (WMO 1986).

In 1988, it was decided to bundle scientific knowledge into an intergovernmental scientific panel to be established under the auspices of the UN when the WMO and the UN Environment Programme (UNEP) established the Intergovernmental Panel on Climate Change (IPCC, see also in the preceding for some of the results). IPCC's main task was to address climate change issues from a policy maker's perspective.

At international climate conferences, the increased climate awareness resulted in international high level discussions among policy makers about potential climate policy measures (Barrett 2005, pp. 355–367). For example, the 1988 climate change conference organized by the Government of Canada recommended the so-called Toronto Target, which stipulated an arbitrary objective of reducing global CO_2 emissions by 20% less than the 1989 levels (Arts 1998, p. 103).

Similarly, the Hamburg Conference of 1988 concluded that "establishment of explicit carbon emission budgets for each country within an international framework of agreement" should be the first international aim and recommended a 30% CO_2 emission reduction (Grubb 1990). This approach was modelled after the successful approach of the Montreal Protocol (Barrett 2005) and the EU's Large Combustion Plant Directive (Grubb 1990). However, Grubb (1990) acknowledged that negotiating such targets and timetables within a global climate policy regime would be an arduous task given that climate change is a global issue that requires a global approach, but which is complicated by large differences between countries in terms of social and economic welfare and the extent to which countries have contributed to GHG emissions.

Nonetheless, since the late 1980s UN-led climate policy negotiations have resulted in several milestones, such as the adoption of the UNFCCC in 1992, the Kyoto Protocol in 1997 and the Cancun Agreements of 2010. These milestones are briefly discussed in the following.

1.3.2 United Nations Framework Convention on Climate Change

In 1989, UNEP and the WMO initiated the preparations for negotiations on a framework convention on climate change. The idea was that this convention would contain basic principles and agreements that could subsequently be worked out in further detail through amendments and protocols. As explained by Grubb (1990), the framework approach has the advantage that it can flexibly deal with evolving scientific knowledge and areas of concern.

During 1990 and 1992, negotiations took place during five sessions of the Intergovernmental Negotiating Committee for a Framework Convention on Climate Change (INC).[8] During these negotiations, it became clear that the positions of industrialised and developing countries strongly differed. Developing countries pointed out that industrialised countries were responsible for most of the historical anthropogenic GHG emissions and argued that the latter group would have to take the lead by adopting quantified emission reduction targets. Most industrialised countries opposed such targets.

Another issue during the negotiations was how international collaboration could be established to support financial and technology transfer to support developing countries in achieving low-emission sustainable development. A key challenge during the negotiations, therefore, was to agree on a global responsibility for the global climate and climatic impacts and how these responsibilities could or should be differentiated between countries based on historical GHG emission patterns and socio-economic welfare levels.[9]

At the fifth session of the INC, the negotiating parties agreed on a compromise text that contained an objective (not legally binding) for industrialised countries to return their GHG emission levels to 1990 by the year 2000, while allowing developing countries to further develop without any quantitative targets. It also became clear that, for the time being, legally binding targets were not feasible.

---

[8] INC was established on 21 December 1990 by the UN General Assembly (Resolution 45/212). It was scheduled that INC would deliver a draft Convention text that would be ready for signature at the 1992 UN Conference on Environment and Development (UNCED, Rio de Janeiro, Brazil, May–June 1992).

[9] This became known under the UNFCCC as the principle of common but differentiated responsibilities.

As explained earlier, the text was adopted on 9 May 1992 at the UN Earth Summit as the UN Framework Convention on Climate Change (UNFCCC 1992). The ultimate objective of the Convention is to achieve a stabilisation of GHG concentrations in the atmosphere "at a level that would prevent dangerous anthropogenic interference with the climate system" (UNFCCC 1992, Article 2). This should be achieved "within a time frame sufficient to allow ecosystems to adapt naturally to climate change, to ensure that food production is not threatened and to enable economic development to proceed in a sustainable manner" (UNFCCC 1992, Article 2).

1.3.3 The Kyoto Protocol

After 1992, negotiations under the UNFCCC continued in the form of annual sessions of the Conference of the Parties or COP.[10] The first session was held in March–April 1995 in Berlin, and here it was concluded that industrialised countries' stabilisation targets as agreed under the Convention were inadequate. This conclusion was based on an analysis of the national GHG emission inventories submitted by 15 industrialised countries (together representing 41% of global GHG emissions), as well as insights from the IPCC Second Assessment Report (IPCC 1995). In its assessment, IPCC (1995) concluded, among other findings, that:

- Atmospheric GHG concentrations had continued to increase;
- Climate had changed since the late nineteenth century; and
- The "balance of evidence suggested a discernible human influence on global climate."

In light of this, the COP initiated a negotiation process (the Berlin Mandate) towards a protocol with quantified emission reduction targets to be achieved within a certain timeframe. The result of this process was the adoption of the *Kyoto Protocol* by the COP in December 1997 (at Kyoto, Japan, UNFCCC 1998). At the heart of the Kyoto Protocol is a GHG emission reduction target for industrialised countries of at least 5% less than 1990 levels within a 5-year commitment period from 2008 to 2012, during which the reduction targets need to be reached. Developing countries were exempted from such a commitment (UNFCCC 1998).

Most of the time during this almost 3-year negotiation process, little progress was made. Again, countries struggled with the principle of "common but differentiated responsibilities" which, in the view of developing countries, implied that

[10] The COP itself was established under the UNFCCC in Article 7 as the supreme body of the Convention with, among others, the task to periodically examine the obligations of the Parties. The first COP in 1995 could only be organised after the entry-into-force of the UNFCCC on 21 March 1994, 90 days after deposit of the 50th instrument of ratification (Portugal, 21 December 1993).

quantified commitments would only apply to industrialised countries. Industrialised countries, on the other hand, doubted the usefulness of a protocol without commitments for, in particular, rapidly developing countries such as China, India, Mexico and Brazil.

Even shortly before the negotiations in Kyoto, the US Congress instructed US negotiators not to agree on a protocol with quantified emission reduction commitments for only industrialised countries and without meaningful participation by the developing country Parties. Such a text would be considered harmful for US economic interests (Fletcher 1997). During the first negotiation week in Kyoto, this position was generally reflected by the interventions by the US delegation. However, there was a change at the beginning of the second negotiation week when US Vice-President Al Gore addressed the COP in person. The US delegation accepted a quantified emission *reduction* commitment, even though developing countries were exempted from such commitments in the Kyoto Protocol.[11] The agreement of the US delegation in Kyoto was never ratified by the US Congress. On the contrary, in March 2001, the newly elected US President George W. Bush decided to withdraw US support from the, in his view, "fatally flawed" text.

Nevertheless, other countries managed to keep the Kyoto process on-going throughout 2001, partly by allowing industrialised countries more room for counting sequestration of carbon in soils and trees and agreeing on more flexible compliance procedures. Although these concessions were criticised for reducing the environmental integrity of the Kyoto Protocol,[12] they secured the support of important industrialised countries such as the EU, Russia, Canada, Japan and Australia.

1.3.4 Towards a Post-2012 Agreement

Immediately after the entry-into-force of the Kyoto Protocol in 2005,[13] a new round of negotiations started about its continuation beyond 2012. By design, the Kyoto Protocol was the most suitable framework for these negotiations as it contained an infrastructure to guide countries through consecutive commitment periods. However, the Kyoto Protocol had not been ratified by all industrialised countries

---

[11] It must be noted that the Kyoto Protocol eventually contained various aspects that the US delegation had introduced in the negotiations, such as the decision that industrialised countries could fulfil their commitments flexibly, i.e., by spreading compliance across a five-year commitment period, by focusing on six GHGs instead of the initial focus on CO_2, and through emissions trading.

[12] For example, because of the concessions, Japan's required emission reduction dropped from 6% less than 1990 levels to a 1% reduction, while Canada could let its GHG emissions increase by 5% (Benedick 2001).

[13] After enough industrialised countries had ratified the Kyoto Protocol so that it represented at least 55% of the total GHG emissions of industrialised countries in 1990.

(e.g., USA) and had kept a strong distinction between industrialised countries with quantified commitments and developing countries without such commitments. Consequently, at the 11th session of the COP in 2005, it was decided to move on with negotiations in two separate tracks: one track would focus on the continuation of the Kyoto Protocol after 2012 (the Ad-hoc Working Group on continuation of the Kyoto Protocol or AWG KP) and the other would focus on long-term cooperative actions by all UNFCCC Parties (the Ad-hoc Working Group on Long-term Cooperative Action or AWG LCA). The latter track was included to bring the USA back into the negotiations, as well as other Parties that did not want to participate in the Kyoto Protocol process. Somehow, it was envisaged that eventually the two tracks would come together into one overall climate policy regime.

In 2007, the COP adopted the Bali Plan of Action (named after the COP's venue of that year), which outlined a process to "reach an agreement on long-term cooperative action up to and beyond 2012" (UNFCCC 2008a, Decision 1/CP.13). A first attempt to reach such an agreement failed in December 2009 when the Copenhagen Climate Conference[14] did not manage to reach consensus on the Copenhagen Accords. The final negotiation text in Copenhagen was, therefore, not legally binding. Instead, the COP took note of the text so that it was formally acknowledged and there was a mandate to follow up on its implementation.

There are two observations here. On one hand, although not legally binding, the text had been drafted by a group of industrialised countries and emerging economies who gave their political support to it. On the other hand, as a consequence of the accords not being a formal COP decision, the several mechanisms proposed in the text (e.g., Copenhagen Green Climate Fund and Technology Mechanism) could not be implemented yet and needed to wait for the formal acceptance of the accords by the COP. The COP session of Cancun (Mexico) in November–December 2010 was the next opportunity for that.

1.3.5 Cancun Agreements: From Targets to Pledges

The *Copenhagen Accords* invited industrialised countries to submit individual or joint economy wide emission targets for the year 2020. In addition, it stated that "non-Annex I Parties to the Convention will implement mitigation actions."[15] On 1 February 2010, less than two month after "Copenhagen," 55 countries had submitted national pledges to cut and limit GHG emissions by 2020. These countries together accounted for 78% of global emissions from energy use. The EU, for instance, pledged a GHG emission reduction for the entire bloc of 20% by the year 2020.

---

[14] COP 15 and fifth meeting of Kyoto Protocol Parties (COP–MOP).

[15] *Non-Annex I* refers to the countries that were not included in the UNFCCC Annexes of Parties with GHG emission stabilisation targets. In practice, the term *non-Annex I Parties* mainly refers to developing countries.

Another important signal came from the so-called BASIC group, with Brazil, South Africa, India and China. On 24 January 2010, this group met in New Delhi and underscored their support to the Copenhagen Accords and announced the following emission reduction pledges (JIQ 2009):

- Brazil: 36% emission reduction below business-as-usual by 2020;
- South Africa: 34% emission reduction below business-as-usual by 2020;
- India: 20% reduction in the carbon intensity by 2020 compared with 2005 levels; and
- China: 45% reduction in the carbon intensity by 2020 compared with 2005 levels.

The US climate position has remained uncertain since Copenhagen. Before December 2009, the Obama administration had proposed a 17% GHG emission reduction less than 2005 levels by the year 2020. This percentage was included in a bill that passed the House of Representative during 2009, as well as in a similar climate bill that was discussed by the Senate. However, the Senate did not pass its bill before Copenhagen, and later in 2010, mid-term elections further reduced the chances of acceptance of the bill.

The Cancun Agreements of December 2010 and the discussions at the COP in Durban (South Africa, November-December 2011) reflected the shift in the negotiations since "Copenhagen" from a top-down architecture where an overarching goal is translated in individual country targets (such as in the Kyoto Protocol) to one in which national pledges should add up to a joint international effort. In line with the Copenhagen Accords, the Cancun Agreements invite countries to formulate national targets (pledges) and agree on international review procedures for these (Kok et al. 2010). These pledges thus involve agreeing on a package of multi-component commitments by each nation based on their domestic circumstances (Reinstein 2004; Stavins 2009). Box 1.2 presents a summary of GHG emission reduction action pledged after December 2009.

Looking at the history of the negotiations, the softer, i.e., more voluntary based, "pledge and review" approach did not come as a surprise. Although Parties such as the Small Island Developing States and the EU emphasised at Cancun the importance of agreeing on an overall long-term climate target (e.g., reducing industrialised country emissions by 25–40% by the year 2020), especially the USA (but also countries such as Canada and Japan) opposed such a top-down approach. Important reasons were the preference to take decisions domestically instead of deriving targets from overarching international goals, and perceived lack of meaningful participation by emerging economies such as India, Brazil and especially China. In addition, several countries made their pledges conditional on the actions proposed by other Parties. For instance, the pledges of several developing countries are conditional on international financial and technical support from industrialised Parties. According to Falkner et al. (2010): "little has thus changed in the way in which the major players in climate politics define their interests."

However, at a webinar organised by the Association of Climate Change Officers (ACCO) on 9 February 2011, US Deputy Special Envoy for climate change,

Jonathan Pershing, explained that the Cancun Agreements' architecture had enabled more countries to become actively involved in a global climate coalition:

> If you ask the Chinese, or in fact the Indians, Indonesians or Mexicans whether they are prepared to undertake legally binding obligations to reduce GHG emissions and enshrine that in international law, the answer so far has been: no. If you ask them if they are prepared to undertake real action and aggressively work to reduce emissions and to make transparent the efforts that they are undertaking, the answer has been yes.

Box 1.2 Summary of GHG emission reduction "pledges"
As a result of the negotiations in Copenhagen and Cancun, more than 100 countries have pledged measures to reduce or limit their GHG emissions. However, further analysis shows that these are insufficient to realise the emission trajectory required to limit climate change to 2°C (Kok et al. 2010; Project Catalyst 2010; Stern 2009; Stern and Taylor 2010; UNEP (2010) and Figueres 2011). Based on presented plans by mid-2011, industrialised countries would jointly reduce GHG emissions by 12–18% less than 1990 levels by 2020. This is lower than the range needed as reported by IPCC (2007) for realising a stabilisation of atmospheric GHG concentrations at 450 parts per million CO_2 equivalent in 2020 (Kok et al. 2010; Den Elzen and Höhne 2008). For instance, Höhne et al. (2011) conclude that presently committed and pledged actions would jointly limit global GHG emissions to 54 gigatonnes per year, whereas this should be 44 gigatonnes (for keeping temperature increase less than the 2°C) and 40 gigatonnes (for a 1.5°C objective) (Höhne et al. 2011). In other words, there is a shortage of 10–14 gigatonnes per year.

1.3.6 Increased Focus on Low-Emission Development Strategies

The Cancun Agreements contain a joint vision for long-term co-operative action between all countries which [16]:

> addresses mitigation, adaptation, finance, technology development and transfer, and capacity-building in a balanced, integrated and comprehensive manner to enhance and achieve the full, effective and sustained implementation of the Convention, now, up to and beyond 2012. (UNFCCC 2010, para 1)

For developing countries in particular, the Cancun Agreements contain a decision that they:

---

[16] The terms low-carbon and low-emission development strategies are used interchangeably in literature sources. In this book, we use the term low-emission development strategies as this indicates more clearly that the focus is on mitigation of a range of greenhouse gases instead of just carbon or carbon dioxide.

> will take nationally appropriate mitigation actions [NAMAs] in the context of sustainable development, supported and enabled by technology, finance and capacity-building, aimed at achieving a deviation in emissions relative to "business as usual" emissions in 2020. (UNFCCC 2010, para 48)

In addition, all Parties are invited to enhance action on adaptation under the Cancun Adaptation Framework, through:

> planning, prioritising and implementing adaptation actions, including projects and programmes, and actions identified in national and sub-national adaptation plans and strategies [with] impact, vulnerability and adaptation assessments, [to] establish a process to enable least developed country Parties to formulate and implement national adaptation plans [NAPs] (UNFCCC 2010, paras 14a–b, 15, and 20e).

As mentioned earlier, developing countries are also encouraged to develop low-carbon development strategies (LCDS) or plans in the context of sustainable development (UNFCCC 2010, para 6). This concept was introduced by a number of Parties during and after the 2009 Climate Conference in Copenhagen. For example, the EU suggested that developing countries prepare low-carbon growth plans from which NAMAs could be formulated (UNFCCC 2009). Another example is the proposal submitted by Republic of Korea, suggesting that:

> …developed country Parties need to provide developing country Parties with a roadmap for low-carbon development which includes appropriate policy tools and necessary support to enable them to pursue greenhouse gas emission reduction and economic development at the same time. (UNFCCC 2009)

A LEDS can be a domestic strategy document that integrates national climate change policy into a broader framework of development (Clapp et al. 2010). Although there is as of yet no clearly established process for LEDS, several studies identify key steps of the process (see, e.g., Clapp et al. 2010; World Bank 2009 and Project Catalyst 2009).

In addition to the decisions on LEDS, NAMA and NAP formulation, the Cancun Agreements:

- Elaborated on how capacity building and financial support to developing countries could be arranged;
- Offered guidelines for reducing GHG emissions through avoided deforestation and reduced land degradation; and
- Established the Technology Mechanism for support of development and transfer of technologies for mitigation and adaptation to developing countries.

As such, the text contained several conceptual agreements between countries, but final decisions on how these concepts would form a new post-2012 climate policy regime were scheduled for future negotiations at COP sessions. This started with the 17th COP in Durban (South Africa, 28 November–9 December 2011) which has taken decisions on the operationalisation of provisions included in the Cancun Agreements (such as the operationalisation of the Adaptation Committee, the Green Climate Fund, and the Technology Mechanism) (UNFCCC 2012).

1.4 The Way Forward

1.4.1 Why Is a Global Climate Policy Approach Necessary?

The fact that GHG emissions are not equally distributed across the globe has led several authors to question whether a global response to climate change is really needed (Victor 2006 and 2007; Prins and Rayner 2007; Haas 2008). According to this argument, given that only around 20 countries in the world are responsible for approximately 80% of global GHG emissions, international negotiations could be considerably simplified if a climate treaty with national legally binding targets was built around the larger emitting countries, without losing much in terms of effectiveness.

Despite these intuitive advantages, it can be questioned if such an approach would be really simpler than the global approach followed thus far under the UNFCCC and the Kyoto Protocol. First of all, the group of countries with highest GHG emissions is still very diverse; not only are major industrialised countries among them but also rapidly developing countries, such as China, India, Mexico and South Africa.[17] As explained earlier, the recent negotiations on a post-2012 climate policy regime have made clear that the latter countries have been unwilling to follow a "target and timetable" approach with national legally binding targets. Depledge and Yamin (2009) argue that the international negotiations under the UNFCCC have de facto also resulted in such a small group with quantified commitments, but with the difference that rapidly developing countries are not part of that group.

Furthermore, a system with emission reduction commitments for only a particular group can lead to "trade leakage," also called "carbon leakage" (Schreuder 2009).[18] Carbon leakage reduces the effectiveness of a geographically limited climate policy regime as companies can move their business to countries without commitments. Therefore, as concluded by Gros and Egenhofer (2010), strong action by a limited number of Parties without comparable action by the other

[17] For example, the Major Economies Forum on Energy and Climate, launched in March 2009 to facilitate a dialogue on energy and climate change among larger developed and developing country economies, contains the following countries: Australia, Brazil, Canada, China, the EU, France, Germany, India, Indonesia, Italy, Japan, Korea, Mexico, Russian Federation, South Africa, USA, and UK.

[18] In the context of climate change policy, Barrett (1997 and 2000) identified two aspects of trade leakage. First, countries with emission reduction commitments would switch from carbon-intensive fossil fuels to fuels with lower carbon content or to renewable energy sources. Because of the lower demand for fossil fuels in these countries, fossil fuel prices will go down, which could create an incentive for countries without commitments to increase their demand for fossil fuels. The GHG emission reduction achieved because of the commitments would thus be offset by increased emissions elsewhere. Second, Barrett (1997 and 2000) and Schreuder (2009) argue that trade leakage can occur if companies decide to shift their production from countries with commitments to countries without commitments.

major emitters is not very useful. Or, as stated by Depledge and Yamin (2009), climate change is inescapably global in nature and a global approach is a fundamental political necessity.

1.4.2 Aligning Climate Policies with Development Policies

As the preceding synopsis has shown, a critical factor during negotiations has been the division between industrialised and developing countries. In the meantime, things have not become easier, on the contrary, the initial distinction between industrialised and developing countries has become increasingly difficult to maintain. During negotiations, it became particularly clear that treating all developing countries as one group does not do justice to the wide diversity among them. For example, in addition to the group of small island state developing countries, which already operated as a negotiation group during the early 1990s, negotiations have increasingly focussed on the position of rapidly growing developing countries and whether these countries would still have to be considered "real" developing countries. According to Depledge and Yamin (2009), the division between industrialised and developing countries is the greatest weakness within the international climate change regime.

Obviously, persuading countries to join an international climate coalition is easier said than done, especially if the coalition's policy package may not be in their national interest. This holds for both industrialised and developing countries, but with respect to the latter countries, Grubb and Patterson (1992) noted that "for many developing countries climate change is a Northern issue, both because it is the North that is primarily interested in it, and because the North created the problem." The challenge has, therefore, been to develop the climate policy framework in such a way that it would be in the interest of all countries, including developing countries' national priorities.

As mentioned at the beginning of this chapter and as is explained in further detail in the next chapter, it has become clear that climate and sustainable development have become increasingly interlinked. For instance, as is discussed in Chap. 5, according to a recent estimate, for making the MDGs in Africa resilient to climate change, 40% more funding will be required during the next 10 years (Anderson 2011). Such increased knowledge of potential impacts and risks of climate change has resulted in an increased focus on the need to reduce countries' vulnerability to climate change impacts, so that the sustainable livelihoods and ecosystem services on which people depend can be protected.[19] This is especially important for least developed countries, which have contributed least to global

[19] Adaptation measures can focus on reduced impacts of climate change on health and social systems and on sectors such as agriculture, biodiversity and ecosystems, production systems, and physical infrastructure, including the energy grid.

GHG emissions, but are likely to be the most vulnerable to climate changes. This equity issue has been acknowledged in the Cancun Agreements that call for the establishment of a process to enable least developed countries to formulate and implement national adaptation plans (UNFCCC 2010, para 15 and 16).

However, it is not only the area of adaptation where climate and development meet. In fact, given the climate and sustainable development urgencies described in this book, there is every reason for countries to realise their economic, social and environmental sustainable development goals with the lowest GHG emissions, and to have this supported under an international climate policy regime. The provisions under the Cancun Agreements, such as formulating NAMAs "in the context of sustainable development" (UNFCCC 2010, para 48), could be important steps in that direction. Furthermore, as is discussed in Chap. 2, the recently updated process of technology needs assessments for climate change (UNDP 2010) takes developing countries' development priorities as a starting point for identifying strategic sectors for achieving climate and sustainable development goals and selecting technologies for mitigation and adaptation with development benefits.

These developments add to past climate negotiation and policy-making processes an increased focus on how to conduct climate mitigation and adaptation actions as an integrated part of domestic sustainable development strategies and policies. Linking climate and development goals in comprehensive strategies, therefore, enables countries to make a change in domestic systems needed to achieve long-term objectives with low emissions and increased climate resilience.

1.4.3 Stimulating Low-Emission Development Through Technology Transfer and Innovation

1.4.3.1 Acceleration of Innovation for Climate and Development

Low-emission and climate-resilient development will involve technology transfer and innovation on a large scale, and we explore these concepts briefly here as they relate to how the climate and development goals of a country may be attained. Further elaboration on technology transfer and innovation is provided in Chaps. 2 and 3. Development literature, such as the Practical Action (2010) report, clearly demonstrates that energy access through low-emission technologies is a major means of addressing poverty alleviation, thus addressing both sustainable development and climate goals. In addition, Practical Action (2010) shows that coping strategies are urgently needed to help countries adapt to climate change impacts and ensure a changing climate does not hamper poor people's access to energy.

Technology transfer is a complex process involving not only equipment or measures but also people. IPCC (2000) defines technology transfer as a set of processes "covering the flows of know-how, experience and equipment, for mitigating and adapting to climate change amongst different stakeholders such as governments, private sector entities, financial institutions, NGOs and research/education institutions."

In line with this, IPCC (2000) describes technology transfer as a process of "learning to understand, utilise and replicate the technology including the ability to decide which technology to transfer and adapt it to local conditions and integrate it with indigenous technologies."

The process of transfer and innovation of the technology or measure into a country system, therefore, involves a dynamic mix of actors, institutions and organisations as well as the hardware and the tacit knowledge needed for success. Furthermore, as will be discussed in the next chapters, technologies can be in different stages of development (e.g., still in the process of research and development, ready for deployment in the market or near commercial application), which will have implications for what the necessary changes are in the country for successful transfer and innovation processes.

The urgency and scale of the climate change problem described in preceding sections means that the normal process of innovation will have to be accelerated, which implies that it is important that there is a strategy and an action plan based on the technologies to be transferred for the country context.

The country context is important here as it determines the current enabling environment for the technologies, including the specific cultural and business habits, language, trust, networks and capacity available for successful transfers. A strategy, as will be described in Chap. 3, may involve a range of activities to overcome barriers to technology diffusion into the market or system. These activities can include enabling policies and measures, networks, market or system support (e.g., quality control), education and training, etc., derived from the accelerating activities required for the specific country and technology characteristics.

The preceding observations are in line with the recent analysis by Jackson (2009) who illustrates how economies in developed and developing countries need to transform to low-emission societies to achieve long-term sustainable development. In Jackson's view, system changes are crucial in this respect, with a central role for innovation.

Within the context of this book, innovation would focus on the process of accelerating behavioural change and adoption of market and non-market (e.g., coping actions) technologies and measures at the research, development, demonstration (RD&D), deployment and diffusion stages for realising both national development objectives and climate policy goals.

1.4.3.2 Assessing Technology Needs for Climate and Development

Transfer of technologies for mitigation and adaptation to developing countries is supported under Article 4.5 of the UNFCCC that calls on developed countries "to promote, facilitate and finance the transfer of or access to environmentally sound technologies and know-how to other Parties, particularly developing country Parties" (UNFCCC 1992, Article 4.5).

As will be explained further in Chap. 2, an important next step was made in 2001 at the COP session in Marrakech (Morocco) with the decision on a Framework for Actions on Article 4.5 and the establishment of an Expert Group on Technology Transfer (EGTT). It also called on Technology Needs Assessments (TNAs) to help developing countries determine mitigation and adaptation priorities. Between 2002 and 2009, 92 developing countries received funding from the Global Environment Facility (GEF) for conducting TNAs.

In 2010, an updated TNA methodology was adopted by the EGTT (UNDP 2010), reflecting lessons learned from earlier TNAs and establishing a closer link to developing country sustainable development priorities as well as formulating overarching strategies for acceleration of innovation based on prioritised technologies.

A TNA is a country-driven process with strong participation of country stakeholders. As mentioned elsewhere in this chapter, TNA starts with a vision on the country's long-term sustainable development, and the main output is an innovation strategy and action plan showing the pathway towards those development and climate goals with the use of low-emission and climate-resilient technologies and measures. Chap. 2 explains in further detail the background for and process of TNA as an element of technology transfer and innovation under the UNFCCC.

After 2008, a new round of TNAs was initiated as part of *Poznań Strategic Programme on Technology Transfer*, adopted by the COP at its session in *Poznań*, Poland (UNFCCC 2008b). This *TNA Project*,[20] which is being implemented by UNEP and for which funding the GEF has provided US$ 9 million starting budget, envisages supporting 35–45 developing countries (UNFCCC 2011a, para 83).

As is explained in further detail in Chap. 4, an important contribution that TNAs could deliver to international climate negotiations is that they provide insights and a process for facilitating and accelerating project, (sub)sector and nationwide transfers and innovation into the country system to facilitate attainment of the country's sustainable development and climate goals. With these insights countries would not only build up knowledge of what would be needed but also how this would need to be done.

1.4.3.3 Progression to a Technology Mechanism

Earlier in this chapter, we have explained how, with respect to technology transfer and innovation, the Copenhagen Accords and Cancun Agreements included a decision:

- To support action on mitigation and adaptation;
- To recognise that technology needs must be nationally determined; and

[20] See for further information: http://tech-action.org/. Accessed 19 September 2011.

- To accelerate action at different stages of the technology cycle, including research, development and demonstration (RD&D), deployment, diffusion and transfer of technologies.

Most importantly, at Cancun, it was decided to transform the EGTT to a Technology Executive Committee (TEC) under the newly established Technology Mechanism. This mechanism will also entail a Climate Technology Centre and Network (CTCN) arm to be established as a virtual entity composed of existing centres (UNFCCC 2011b). The objective of the Technology Mechanism is to become the centrepiece of an agreed outcome on technology and development transfer under the Convention (UNFCCC 2011b). The TEC could be considered as the mechanism's "political arm" with the objective to support developing countries in developing and enhancing capacity for development and transfer of both soft and hard technologies for mitigation and adaptation, as well as in getting access to private and public investment funding.

The CTCN has a more operational focus with the objective to support developing countries, at their request, with operational modalities (UNFCCC 2011b), such as:

- Offering or guiding countries towards products that they can use for making technology choices and policy preparation and implementation, such as analytical tools, information tools and good practices;
- Offering or providing access to *services*, such as training, advice and expert teams; and
- Establishing or facilitating access to international *partnerships*, such as twinning, and North–South, South–South, South–South–North co-operation.

The CTCN could also facilitate prompt action on deployment of technologies in developing countries (UNFCCC 2011b, p. 11, and UNFCCC 2010 para 123a.iii). For example, a country that has identified prioritised technologies or measures for mitigation or adaptation and formulated projects or programmes for their implementation could request advice from the CTCN through the network about the feasibility of projects or programmes. Such advice or even validation could increase the chance of their successful implementation within the country. It would also establish contacts between country stakeholders and experts "out there" but available for consultation through the CTCN. These interlinkages are further discussed in Chap. 4.

1.5 Meeting the Challenge: An Integrated Approach to Sustainable Development and Climate Change for Global Concerted Action

In this chapter, we have discussed the challenge for policy makers to place climate negotiations in the wider context of sustainability, equity and social change to address the concerns of developing countries and to aid progress with the climate change problem at the scale and with the urgency needed.

It has been argued that climate and sustainable policy making have become strongly interlinked: Climatic changes require adaptation by individual countries to achieve development goals, whereas in light of the climate urgencies, there is every reason to achieve development goals with the lowest GHG emissions possible. This approach also enables development benefits and mitigation and adaptation benefits to be maximised for efficient allocation of resources.

The GHG emission reduction actions promised by countries thus far under the Copenhagen Accords and Cancun Agreements are insufficient to limit the increase in global temperature to less than 2°C, and we are only at 60% of what is required (Figueres 2011).

However, the latest climate negotiations have resulted in the building blocks for addressing our challenges. The concepts of Low-Emission Development Strategies, Nationally Appropriate Mitigation Actions, National Adaptation Programmes and the updated Technology Needs Assessment process are examples of such building blocks.

It shows that there is increasing insight that for a global climate coalition all countries need to work together and that can only be achieved if countries' mitigation and adaptation actions are embedded in national preferences, including developing countries' needs for adaptation and mitigation actions in support of achieving the Millennium Development Goals.

These issues are explored in the next chapters. Chap. 2 explains how countries could align their sustainable development with technology transfer needs for climate change mitigation and adaptation (challenge 2). Chapter 3 discusses the process of identifying actions for accelerating innovation for climate and development goals and how this could be formulated as a strategy with action plans (challenge 3). Chapter 4 addresses mechanisms by which current directions in international climate policy making can be integrated and extended for coherent and efficient support in order that technology transfers achieve the climate and sustainable development goals (challenge 4). Chapter 5 discusses the finance needs for low-emission and climate-resilient development, both in terms of the scale of funding requirements and tailoring the types and sources of funding. The focus will not only be on technologies and measures but also on financing the support actions identified for accelerated innovation into the country, as well as for energy access for the poor (challenge 5).

The book will conclude with a short overview of suggested solutions for these challenges in Chap. 6.

References

ACCO (2011). *Cancun de-brief: Reflections on COP-16 in cancun and implications for U.S. Businesses*. Webcast organized by the association of climate change officers (ACCO) on 9 February 2011. Retrieved February 9, 2011 from http://www.accoonline.org/event_archives.html#Cancun De-brief: Reflections on COP-16 in cancun and implications for U.S. Businesses.

AGECC (2010). *Energy for a sustainable future*. The Secretary-General's Advisory Group on energy and climate change (AGECC). Summary report and recommendations, 28 April 2010, New York, USA. Retrieved September 23, 2011 from http://www.un.org/wcm/webdav/site/climatechange/shared/Documents/AGECC%20summary%20report[1].pdf.

Anderson, S. (2011). *Climate change and poverty reduction*. Climate & Development Knowledge Network (CDKN). Policy Brief, August 2011. Retrieved August 9, 2011 from http://cdkn.org/wp-content/uploads/2011/08/CDKN_poverty-reduction_FINAL.pdf.

Arts, B. (1998). *The political influence of global NGOs. Case studies on the climate and biodiversity conventions*. Utrecht, The Netherlands. International Books.

Barrett, S. (1997). The strategy of trade sanctions in international environmental agreements. *Resource and Energy Economics, 19*, 345–361.

Barrett, S. (2000). Trade and environment: Local versus multilateral reforms. *Environment and Development Economics, 5*, 349–359.

Barrett, S. (2005). *Environment and Statecraft, the strategy of environmental treaty making*. Oxford: Oxford University Press.

Benedick, R. E. (2001). Striking a new deal on climate change. *Issues in Science and Technology, 18*(1), 71.

Berkeley Earth Surface Temperature Project (2011). *Preliminary Findings*. Retrieved April, 26 2012 from http://berkeleyearth.org/findings.

Clapp, C., Briner, G., & Karousakis, K. (2010). *Low-emission development strategies (LEDS): Technical, Institutional and policy lessons*. Paris: OECD, IEA. Retrieved September 23, 2011 from http://www.oecd.org/dataoecd/32/58/46553489.pdf.

Den Elzen, M., & Höhne, N. (2008). Reductions of greenhouse gas emissions in Annex I and non-Annex I countries for meeting concentration stabilisation targets. *Climatic Change, 91*, 249–274.

Depledge, J., & Yamin, F. (2009). The global climate-change regime: A defence. In D. Helm, C. Hepburn, (Eds.), *The economics and politics of climate change* (1st ed.). Oxford: Oxford University Press.

European Commission (2011). *A Roadmap for moving to a competitive low carbon economy in 2050*. Communication from the Commission to the European Parliament, the Council, the European Economic and Social Committee and the Committee of the Regions. *COM/2011/0112 final*/. Retrieved September 22, 2011 from http://ec.europa.eu/clima/documentation/roadmap/index_en.htm

Falkner, R., Stephan, H., & Vogler, J. (2010). International climate policy after Copenhagen: Towards a 'building blocks' approach. *Global Policy, 1*(3), 252–262.

Figueres, C. (2011). *Statement by Christiana Figueres at the high-level inter-parliamentary debate on climate change*. Brussels, Belgium, 19 April 2011. Retrieved September 17, 2011 from http://unfccc.int/files/press/statements/application/pdf/110419_speech_inter-parliament_brussels.pdf

Fletcher, S. R. (1997). *Global climate change treaty: summary of the Kyoto Protocol*. Congressional research service report for congress. 98–2 ENR. The National Council for Science and the Environment, Washington, D.C., USA.

Gore, A. (2006). *An inconvenient truth: The planetary emergency of global warming and what we can do about it*. Emmaus, Pennsylvania, USA: Rodale Press.

Gros, D., & Egenhofer, C. (2010). *Climate change and trade: Taxing carbon at the border?* CEPS Paperback. Brussels, Belgium. Retrieved August 18, 2011 from http://www.ceps.eu/system/files/article/2009/08/Climate%20Change%20and%20Trade.pdf

Grubb, M. (1990). The greenhouse effect: Negotiating targets. *International Affairs, 66*,(1), 67–89 (Royal Institute of international Affairs).

Grubb, M., & Patterson, M. (1992). The international politics of climate change. *International Affairs, 68*, 293–310.

Haas, P. M. (2008). Climate change governance after Bali. *Global Environmental Politics, 8*(3), 1–7.

Höhne, N., Hare, B., Schaeffer, M., Chen, C., Vieweg, M., & Moltmann, S. (2011). *Emissions and CO_2 concentrations at record highs: Developed countries ambitions stalled while developing countries gearing up to act*. Climate Action Tracker Update, 16 June 2011. Retrieved June 28, 2011 from http://www.climateactiontracker.org/CAT_update_Bonn_2011-06-16.pdf

IARU. (2009). *Climate change: Global risks, challenges and decisions*. International alliance of Research Universities. Retrieved April 29, 2011 from http://www.climatecongress.ku.dk/

IEA (2009). *World energy outlook 2008*. International Energy Agency. Paris: OECD/IEA. Retrieved May 28, 2011 from http://www.iea.org/weo/

IEA (2011). *World Energy Outlook 2011*. International Energy Agency. Paris: OECD/IEA.

IPCC (1995). *Climate change 1995: Synthesis report contribution of working groups I, II and III to the second assessment report of the Intergovernmental Panel on Climate Change*. Geneva, Switzerland. Retrieved May 27, 2011 from http://www.ipcc.ch/pdf/climate-changes-1995/ipcc-2nd-assessment/2nd-assessment-en.pdf

IPCC (2000). *Methodological and technological issues in technology transfer—a special report of IPCC working group III*. Intergovernmental Panel on Climate Change. Geneva, Switzerland. Retrieved September 21, 2011 from http://www.ipcc.ch/pdf/special-reports/spm/srtt-en.pdf

IPCC (2007). Climate change 2007: Synthesis report. In R. K. Pachauri & A. Reisinger (Eds.), *Contribution of working groups I, II and III to the fourth assessment report of the Intergovernmental Panel on Climate Change*. Geneva, Switzerland: IPCC.

Jackson, T. (2009). *Prosperity without growth: Economics for a finite planet* (1st ed.). London: Earthscan.

JIQ (2009). Uncertainty remains after Copenhagen. *Joint Implementation Quarterly, 15*(4), 2–3.

Kok, M. T. J., Lüdeke, M. K. B., Sterzel, T., Lucas, P. L., Walter, C., Janssen, P., & De Soysa, I. (2010). *Quantitative analysis of patterns of vulnerability to global environmental change*, Netherlands Environmental Assessment Agency (PBL), Den Haag/Bilthoven, the Netherlands. Retrieved September 10, 2011 from http://www.pbl.nl/bibliotheek/rapporten/550025005.pdf

LDC Expert Group (2009). *National adaptation programmes of action: Overview of preparation, design of implementation strategies and submission of revised project list and profiles*. UNFCCC. Retrieved September 10, 2011 from http://unfccc.int/resource/docs/publications/ldc_tp2009.pdf

Meinhausen, M., Meinhausen, N., Hare, W., Raper, S. C. B., Frieler, K., Knutti, R., et al. (2009). Greenhouse-gas emission targets for limiting global warming to 2°C. *Nature, 458*, 1158–1162.

Muir Russell (2010). *The Independent Climate Change E-Mails Review*. Retrieved April 26, 2012 from http://www.cce-review/pdf/FINAL%20REPORT.pdf.

New, M., Liverman, D., Schroder, H., & Anderson, K. (2011). Four degrees and beyond: The potential for a global temperature increase of four degrees and its implications. *Philosophical Transactions of the Royal Society, 369*, 6–19.

Practical Action (2010). *Poor people's energy outlook 2010*. Rugby, UK. Retrieved September 15, 2011 from http://practicalaction.org/ppeo2010

Prins, G., & Rayner, S. (2007). *The wrong trousers: Radically rethinking climate policy*. Joint discussion paper of the James Martin Institute for Science and Civilization, Oxford, University of Oxford, and the MacKinder Centre for the Study of Long-wave events, London, London School of Economies and Political Science.

Project Catalyst (2009). *Low carbon growth plans: Advancing good practice*. Working draft 7 August 2009. ClimateWorks and European Climate Foundation. Retrieved September 23, 2011 from http://www.project-catalyst.info/index.php?option=com_content&view=article&id=8&Itemid=12

Project Catalyst (2010). *Taking stock—the emission levels implied by the pledges to the Copenhagen Accord*. European Climate Foundation and ClimateWorks Foundation Briefing paper, February 2010. Retrieved September 22, 2011 from http://www.climateworks.org/download/?id=e63a5561-cff7-4213-9ebf-19297be70623

Reinstein, R. A. (2004). A possible way forward on climate change. *Mitigation and Adaptation Strategies for Global Change, 9*, 295–309.

Rogers, P. (2010). *Losing control: Global security in the 21st century* (3rd ed.). London: Pluto Press.

Schreuder, Y. (2009). *The corporate greenhouse: Climate change policy in a globalizing world.* London: Zed Books Ltd.

Stavins, R. N. (2009). *A portfolio of domestic commitments: Implementing common but differentiating responsibilities*. Harvard: Harvard project on international climate agreements viewpoints.

Stern, N. (2009). *Action and ambition for a global deal in Copenhagen.* Centre for Climate Change Economics and Policy. Grantham Research Institute on Climate Change and the Environment policy paper, December 2009. Retrieved September 22, 2011 from http://www.unep.org/pdf/climatechange/actionandambitionforglobaldealincopenhagen.pdf

Stern, N., & Taylor, C. (2010). *What do the appendices to the Copenhagen accord tell us about global greenhouse gas emissions and the prospects for avoiding a rise in global average temperature of more than 2°C?* Centre for Climate Change Economics and Policy Grantham Research Institute on Climate Change and the Environment policy paper, March 2010. Retrieved September 22, 2011 from http://www.unep.org/PDF/PressReleases/Accord_targets_paper.pdf

UN (1992). *UN Conference on Environment and Development (1992).* Retrieved September 5, 2011 from http://www.un.org/geninfo/bp/enviro.html

UNDP (2010). *Handbook for conducting technology needs assessment for climate change.* Retrieved September 23, 2011 from http://unfccc.int/ttclear/pdf/TNA%20HANDBOOK%20EN%2020101115.pdf

UNEP (2010). *The emissions gap report: Are the Copenhagen Accord pledges sufficient to limit global warming to 2°C or 1.5°C?—A preliminary assessment.* Retrieved September 23, 2011 from http://www.unep.org/publications/ebooks/emissionsgapreport/pdfs/EMISSION_GAP_REPORT_%20HIGHRES.pdf

UNFCCC (no date). *GHG data from UNFCCC.* Retrieved July 16, 2011 from http://unfccc.int/ghg_data/ghg_data_unfccc/items/4146.php

UNFCCC (1992). *United Nations framework convention on climate change*, FCCC/INFORMAL/84. Retrieved May 23, 2011 from http://unfccc.int/resource/docs/convkp/conveng.pdf

UNFCCC (1998). *Kyoto Protocol to the United Nations Framework Convention on Climate Change.* Retrieved May 23, 2011 from http://unfccc.int/resource/docs/convkp/kpeng.pdf

UNFCCC (2008a). *Report of the Conference of the Parties on its thirteenth session*, held in Bali from 3 to 15 December 2007. FCCC/CP/2007/6/Add.1. Retrieved July 17, 2011 from http://unfccc.int/resource/docs/2007/cop13/eng/06a01.pdf#page=3

UNFCCC (2008b). *Development and transfer of technologies.* Decision 2/CP.14, FCCC/CP/2008/7/Add.1. Retrieved September 21, 2011 from http://unfccc.int/resource/docs/2008/cop14/eng/07a01.pdf#page=3

UNFCCC (2009). *Ideas and proposals on the elements contained in paragraph 1 of the Bali Action Plan.* Submissions from parties, part II, FCCC/AWGLCA/2009/MISC.4 (Part II), 19 May 2009. Retrieved September 23, 2011 from http://unfccc.int/resource/docs/2009/awglca6/eng/misc04p01.pdf

UNFCCC (2010). *The Cancun Agreements: Outcome of the work of the Ad hoc Working Group on Long-term Cooperative Action under the Convention.* Decision 1/CP.16. FCCC/CP/2010/7/Add.1. Retrieved July 17, 2011 from http://unfccc.int/resource/docs/2010/cop16/eng/07a01.pdf#page=2

UNFCCC (2011a). *Enhancing the implementation of the results of TNAs.* Background paper II for UNFCCC workshop on technology needs assessments, Bonn, Germany, 1–2 June 2011. Retrieved June 3, 2011 from http://unfccc.int/ttclear/pdf/Workshops/Bonn%202011/BG%20paper%20II-implementing%20TNAs.pdf

UNFCCC (2011b). *Interlinkages between technology needs assessments and national and international climate policy making processes.* Background paper III for UNFCCC workshop on technology needs assessments, Bonn, Germany, 1–2 June 2011. Retrieved June 3, 2011

from http://unfccc.int/ttclear/pdf/Workshops/Bonn%202011/Background%20Paper%20III%2025%20May%20Final.pdf
UNFCCC (2012). *Decisions adopted by the Conference of the Parties.* Report of the Conference of the Parties on its seventeenth session, held in Durban from 28 November to 11 December 2011, Addendum Part Two: Action taken by the Conference of the Parties at its seventeenth session, FCCC/CP/2011/9/Add.1. Retrieved April 26, 2012 from http://unfccc.int/resource/docs/2011/cop17/eng/09a01.pdf
UNFCCC-SBI (2011). *National greenhouse gas inventory data for the period 1990–2009.* Report on national greenhouse gas inventory data from Parties included in Annex I to the Convention for the period 1990–2009, FCCC/SBI/2011/9. Retrieved December 10, 2011 from http://unfccc.int/resource/docs/2011/sbi/eng/09.pdf
Victor, D. G. (2006). Toward effective international cooperation on climate change: Numbers, interests and institutions. *Global Environmental Politics, 6*(3), 90–103.
Victor, D. G. (2007). Fragmented carbon markets and reluctant nations: Implications for the design of effective architectures. In J. Aldy & R. N. Stavins (Eds.), *Architectures for agreement: Addressing global climate change in the post-Kyoto world, chapter 4*. Cambridge: Cambridge University Press.
World Bank (2009). *Low-carbon growth studies—Getting started. Experience from six countries.* Retrieved September 23, 2011 from http://www.esmap.org/esmap/node/362
WMO (1979). *Proceedings of the world climate conference: A conference of experts on climate and mankind.* Geneva, 12–32 February 1979. WMO-No. 537. ISBN 92-63-10537-5.
WMO (1986). *Report of the international conference on the assessment of the role of carbon dioxide and of other greenhouse gases in climate variations and associated impacts*, Villach, Austria, 9–15 October 1985, WMO No.661.

Chapter 2
Challenge 2: Integrating Sustainable Development and Technology Transfer Needs

Abstract This chapter addresses the challenge of how sustainable development and climate goals could be aligned through technology transfers. It explains how this is an opportunity for maximising the benefits from investments at a time of constrained resources but large-scale low-emission technology transfers for mitigation and adaptation. This could facilitate developing countries' efforts to achieve Millennium Development Goals. The chapter reviews the development of technology transfer under the UNFCCC and the updated Technology Needs Assessment (TNA) process. How technologies and measures can be identified for achieving both climate and development goals using the new TNA process involving developing country stakeholders is described as a first stage in developing strategies and action plans for large-scale sector transformations.

2.1 Challenge 2: Introduction

Chapter 1 showed that the world is facing a serious challenge to bend global greenhouse gas (GHG) emission trends to pathways that limit the risks of irreversible damage to the world's ecosystems. International climate negotiations under the United Nations Framework Convention on Climate Change (UNFCCC) have thus far not resulted in a global coalition that is capable of addressing this challenge. Negotiations have long focussed on agreeing on quantitative national GHG emission reduction commitments for industrialised countries, and attempts to broaden this coalition with developing countries have not been successful.

Since the failure of the 2009 Copenhagen Climate Conference, a different approach has been taken and further encouraged by the 2010 Cancun Agreements, whereby countries can report under the UNFCCC what they are planning to do in terms of GHG emission reductions ('pledges'). However, as explained by

W. van der Gaast and K. Begg, *Challenges and Solutions for Climate Change*, Green Energy and Technology, DOI: 10.1007/978-1-84996-399-2_2,

UNFCCC Executive Secretary Christiana Figueres, by mid-2011, the pledges submitted by countries in total amounted to only 60% of what is needed to reach the 2°C target (Figueres 2011). Scaling up the efforts towards achieving climate goals will, therefore, require a 'green industrial revolution'. Transfers of low-emission technologies will be required for this 'revolution' and for developing countries to be engaged, these transfers will have to be effected through alignment of climate actions with countries' sustainable development objectives. This brings us to the second challenge for this book, which is addressed in this chapter:

To meet both countries' sustainable development and technology transfer needs by selecting technologies or measures for climate change mitigation and adaptation based on countries' sustainable development and climate goals.

In the past, several climate-friendly activities have already been carried out in support of developing countries' sustainable development, but most experience so far has been at the project level rather than sector or subsector level. Scarce resources to achieve this challenge will, therefore, need to be allocated to maximise the benefits from investments. We discuss this later on in this chapter.

We will first discuss the engagement of developing countries through sustainable development and poverty alleviation, before moving to how technologies and measures for achieving climate and development goals can be identified.

2.2 Engagement of Developing Countries Through Sustainable Development and Poverty Alleviation

To form a full global climate coalition to take on this challenge, integration of sustainable development priorities of developing countries into the planning of a *green revolution* will be important. The key reason for this is that developing countries are not interested just in climate change as was seen under the climate negotiations, but mainly focus on their country development and poverty alleviation. Therefore, full engagement of developing countries will require that their sustainable development priorities are addressed as well as action on climate change.

Focussing on getting the GHG emission levels lowered on the scale required should not mean that development issues such as energy access for the poor and equity concerns will become less relevant. On the contrary, not integrating climate policy actions and the processes required for sustainable development implies a risk that the climate solutions will not be lasting and that opportunities and resources are wasted. Climate actions delivering sustainable development benefits are likely to be successful, as stakeholders know that these actions are in support of their own interests and urgencies. This has been clearly demonstrated by literature on development co-operation success and failure stories. If we want successful climate policy action, people in the countries need to be properly engaged in the process.

In addition to the argument that aligning climate change mitigation action with sustainable development would enhance the probability that both objectives will be successfully achieved, the relationship between climate and development also becomes clear when looking at the possible consequences of climate change for sustainable livelihoods in developing countries. For instance, as explained by Anderson (2011), climate change makes achieving and sustaining development goals increasingly difficult, including increased problems for poor population groups to 'climb out and stay out of poverty'. He, therefore, recommends that policy instruments for poverty reduction and for adaptation to climate change impacts need to be integrated and that 'identifying how mitigation strategies can also reduce poverty and support adaptation is an important part of climate-resilient development'.

This is in line with Art. 4.9 of the UNFCCC (1992) which specifically addresses the case of the Least Developed Countries (LDCs):

> Parties shall take full account of the specific needs and special situations of the Least Developed Countries in their actions with regard to funding and transfer of technology.

As explained in Chap. 1, this focus on LDCs is an attempt to embed equity into the Convention as LDCs have contributed least to the emission of GHGs but are among the most at risk from climate change effects either because of the vulnerability of the region and/or low capacity to adapt to the changes.

In other words, the way the world deals with climate change today will have a direct bearing on the development prospects of large population groups in developing countries. Enabling adaptation to reduce vulnerability is a key part of the integration process for development and climate goals.

This has been recognised for some time with the LDC Expert Group of the UNFCCC (LDC Expert Group 2009) and the OECD Development Assistance Committee (OECD 2009) concerned with integrating development and adaptation. If these could also be combined with climate mitigation action, then there could be the elusive 'triple win'.

The question then becomes how climate change mitigation and adaptation measures can be identified in light of a developing country's sustainable development objectives and successfully implemented on the scale necessary to avoid serious climate change.

Of course, what is an urgent need in one country is likely to be different from that in another country. Therefore, the choice and design of actions need to be local to ensure that solutions are in line with local priorities. Nonetheless, some overall conclusions can be drawn from recent research conducted in developing country regions. For example, the recent *Poor People's Energy Outlook* by Practical Action (2010) presents some clear examples of current urgencies in developing countries.

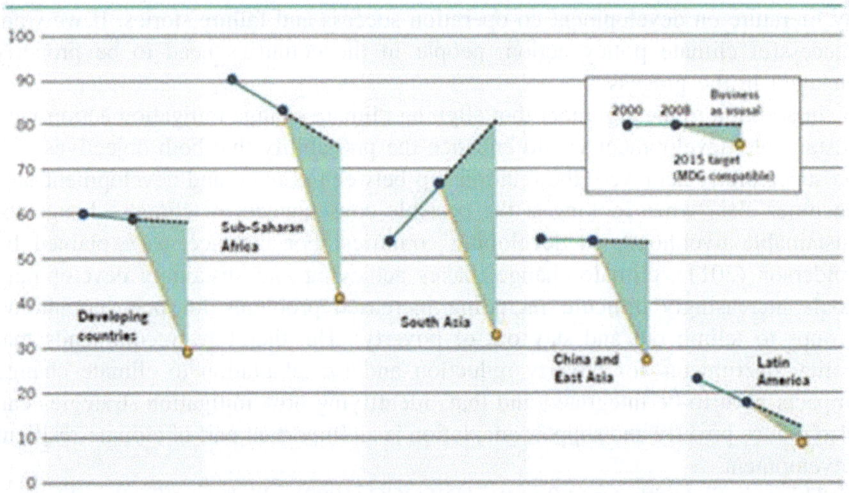

Fig. 2.1 Percentage of people without access to modern fuels for cooking; progress towards the MDG-compatible target (Practical Action 2010, taken from IEA 2002; Legros et al. 2009) (The *curves* show for each region, as well as for all developing countries on average, how access to modern fuel technologies for cooking has been increased or decreased. The *triangles* show the difference between the business-as-usual trends and the Millennium Development Goals for these regions)

The document clearly shows that in several developing country regions, still many serious problems exist with energy access for large groups of people. For instance, they conclude that one and a half billion of the world's population have no access to electricity and that three billion people rely on traditional biomass and coal for cooking. This energy access problem is an important reason why several regions are still far away from reaching the Millennium Development Goals (MDGs). Figure 2.1 illustrates this by showing for different regions what progress has been made between 2000 and 2008 in terms of reducing the number of people without access to modern fuels for cooking, but also by indicating how big the distance is between the present business-as-usual trends and the MDGs.

The figure shows that in Sub-Saharan Africa, access has improved (from 90% to around 75% of people without access to modern fuel cooking technologies), but that for reaching the MDG of 40% access, still much work remains to be done. For South Asian developing countries, the percentage of people without access to modern technologies has even increased, resulting in an almost 50%-point deviation from the MDG. For all developing countries, the average percentage of people without access is now <60%, whereas the MDG is <30% without access.

Without energy access, people would spend a lot of time on collecting biomass for daily cooking and heating services, time which they could have spent better on education, other types of labour, etc. Lack of access to modern energy, therefore, keeps these people in poverty.

In the next section, the role of technology transfer of low-emission technologies to meet both climate and development needs is discussed.

2.3 The Role of Technology Transfer

2.3.1 Why is Technology Important?

Technology transfer can be a powerful solution for simultaneously addressing the climate change and development challenges described earlier. This was recognised in Art. 4.5 of the UNFCCC. Also, as explained in Chap. 1, there are increasing insights that meeting a growing global energy demand with improved energy access for the poor can only go hand in hand with low GHG emission pathways. The recognition that these pathways involve rapid innovation of low-emission technologies has moved technology development and transfer to the heart of the climate negotiations and development debate.

Although technology transfer was discussed at succeeding sessions of the UNFCCC Conference of the Parties (COP) within the context of the Convention's Art. 4.5, it was not until 2001 that significant change occurred. At the seventh session of the COP (or COP 7), held in Marrakech (Morocco), a decision was made on a *Development and Transfer of Technologies* and the *Expert Group on Technology Transfer* (EGTT) was set up to facilitate transfers (UNFCCC 2002). The key themes of the framework adopted by the EGTT were (UNFCCC 2002):

- Assessment of technology needs;
- Technology information: technical and other information;
- Enabling environments: how to solve policy and legal barriers;
- Capacity building: identifying country needs; and
- Mechanisms: co-ordination of process and formulation of projects.

New areas introduced were:

- Innovative options for financing technology transfers; and
- Technologies for adaptation.

In 2009, the EGTT estimated the additional financing needs for low-emission technologies in developing countries at US$ 105–402 billion per year (which is 40–60% of global climate technology finance needs) (EGTT 2009). It was also concluded that:

not all countries have the technologies needed or the ability to innovate new technologies to mitigate and adapt to climate change. Those countries that are lacking in the technologies or capacity, mainly the developing countries, need to be helped not merely to adopt the existing environmentally friendly technologies but also to develop the capacity to innovate new technologies and practices in co-operation with others.

As is discussed in detail in Chap. 3, it is important to underline that identification of technologies and possibly implementing them in projects may not be enough to initiate a system change for widespread technology innovation in a country. Although identification of technologies is an important step in low-emission and climate-resilient development, overarching strategies will be required to make sure that the technologies diffuse well within countries' systems or markets. The strategies may include activities such as organisational/institutional behavioural change, system-supporting services (e.g., finance and legal support), network creation and support, skills training, international co-operation and intellectual property rights and corresponding policies and measures.

This has been made clear by EGTT (2009, p. 11) as follows:

Technology transfer includes not merely transfer of hardware but also of best practices, information and improvement of human skills, especially those possessed by specialized professionals and engineers. The acquisition and absorption of foreign technologies, and their further development, are complex processes that demand considerable knowledge and efforts on the part of those that acquire them. It is the capacity of the countries and the enabling environment in those countries that will enable them to change to a low-carbon economy.

2.3.2 Negotiation Context for Technology Needs Assessments

COP 7 encouraged 'developing countries…to undertake assessments of country-specific technology needs, subject to the provision of resources, as appropriate to country-specific circumstances' (UNFCCC 2002). These Technology Needs Assessments (TNAs) were defined as:

a set of country-driven activities that identify and determine the mitigation and adaptation technology priorities of Parties other than developed country Parties…particularly developing country Parties (UNFCCC 2002).

To support countries in conducting TNAs, the United Nations Development Programme (UNDP) developed a TNA handbook.[1] After 2002, 92 developing countries received funding from the Global Environment Facility (GEF) for conducting TNAs.[2] About 78 of these assessments have been supported by UNDP, and 14 by the United Nations Environment Programme (UNEP).

---

[1] This was done in collaboration with the Climate Technology Initiative (CTI), EGTT and the UNFCCC Secretariat (UNDP 2010, p. 4).

[2] Based on 68 TNA case studies, the UNFCCC secretariat prepared *the Second Synthesis Report on Technology Needs Identified by Parties not included in Annex I to the Convention* (UNFCCC 2009).

At COP 13, held at Bali (Indonesia) in December 2007, the importance of technology transfer under the Convention was further emphasised as a building block for a future climate policy regime (UNFCCC 2008a). The GEF was requested to elaborate a strategic programme to scale up the level of investments for technology transfer to help developing countries assess their needs for environmentally sound technologies. Part of this programme would be support for countries to conduct TNAs or update earlier assessments. The programme was adopted at COP 14 (Poznań, Poland, December 2008) as the *Poznań Strategic Programme on Technology Transfer* (UNFCCC 2008b). As explained in Chap. 1, it envisaged supporting 35–45 developing countries to prepare or update TNAs and formulate technology action plans as TNA output. For the resulting *TNA Project*, which is being implemented by UNEP, the GEF has provided funding of US$ 9 million (UNFCCC 2011).[3]

To support the continued TNA activities, COP 13 requested the UNFCCC Secretariat 'in collaboration with the EGTT, United Nations Development Program (UNDP), United Nations Environment Program (UNEP) and Climate Technology Initiative (CTI), **to update the handbook for conducting technology needs assessments**' (UNFCCC 2008a, bold added).

The updated handbook with the new TNA process was endorsed by the EGTT in November 2010 (UNDP 2010).

The GEF/UNEP TNA Project started in 2009 using an advanced version of the updated TNA handbook with a first round in which 15 developing countries participated (see Box 2.1).[4] These countries are supported by three regional centres through a help-desk facility, as well as by regional training workshops. The output from the TNA conducted under this project is expected to be a Technology Action Plan (TAP) which could be equivalent to a technology-level strategy under the new TNA process.

Box 2.1 TNAs under the Poznań Strategic Programme on Technology Transfer

The objectives of the GEF/UNEP TNA Project are:

- To identify and prioritise through country-driven participatory processes, technologies that can contribute to mitigation and adaptation goals of the participant countries, while meeting their national sustainable development goals and priorities.
- To identify barriers hindering the acquisition, deployment and diffusion of prioritised technologies.

---

[3] See for further details http://tech-action.org/. Accessed 19 September 2011.

[4] These countries were Senegal, Kenya, Mali, Morocco, Cote d'Ivoire from Africa, Cambodia, Indonesia, Viet Nam, Bangladesh and Thailand from Asia, Costa Rica, Guatemala and Peru from Latin America, and Georgia from the Commonwealth of Independent States.

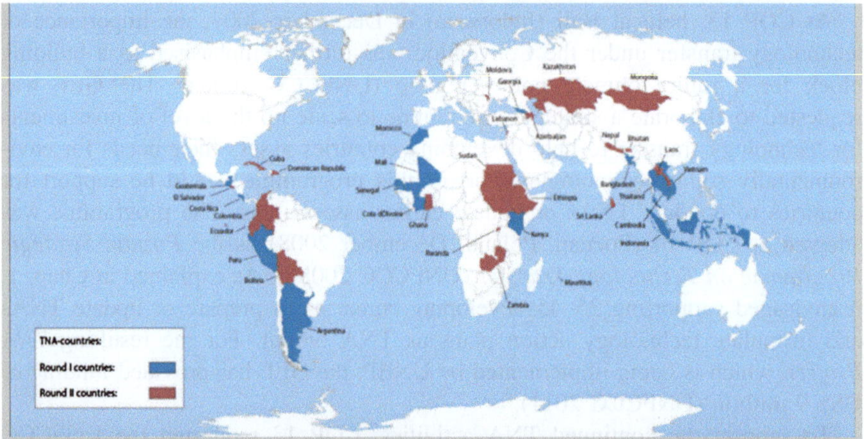

Fig. 2.2 Overview of TNA Project round 1 and round 2 countries (Agbemabiese and Painuly 2011)

- To develop Technology Action Plans specifying activities and enabling frameworks to overcome the barriers and facilitate the transfer, adoption and diffusion of selected technologies in the participant countries.

An initial round of 15 developing countries from Africa, Asia, Latin America and the Caribbean and Europe have been supported in the first round since early 2010. An additional 21 countries were selected in the last quarter of 2010 (see Fig. 2.2).

2.3.3 Starting the Process

The provisions that have been developed under the UNFCCC and included in the Cancun Agreements, such as low-carbon development strategies (LCDS) and TNA, as well as nationally appropriate mitigation actions (NAMA) and national adaptation plans (NAP), all enable a developing country-specific identification of low-emission and climate-resilient development measures. As introduced in Chap. 1, of these provisions, LCDS and TNA aim at formulating sector and/or national strategies, whereas NAMAs and NAPs are actions that could result from such strategies and the policies and measures derived from these.

In this chapter, the process of identifying mitigation and adaptation technologies and measures in the light of a country's national development priorities will be illustrated with help of the TNA process.

As explained earlier, The TNA concept was initiated in 2001 as part of the Marrakech Accords' Technology Framework (UNFCCC 2002) and applied by more than 90 developing countries between 2001 and 2007. Based on experience gathered from these TNAs, the methodology was updated and extended from a mainly technology project orientation to a more technology innovation strategy point of view (UNDP 2010). So far, TNA is the *only* detailed methodology that has been adopted under the UNFCCC compared with the relatively new concepts of LCDS, NAMA and NAP.[5] Later in Chap. 4, links between the TNA approach and LCDS, NAMA and NAP will be explored.

The overall TNA process involves the formulation of strategies and action plans for enabling a change to low-emission sustainable development. It is considered in two main stages:

- The first stage is identification of the technologies or measures for a country that could be used to reduce GHG emissions and climate change vulnerability at the same time as delivering the required sustainable development benefits.
- The second stage is the identification of activities to accelerate the innovation into the country system by identifying actions for overcoming barriers and then formulating them into a strategy and action plan at the technology, (sub)sector or national level.

This chapter continues with an introduction and brief summary of the new TNA process in the updated TNA handbook (UNDP 2010). Comparisons are made with previous TNA exercises, and the key differences between the new and the previous approach are then highlighted. This is followed by a more detailed illustration of the first of the two TNA stages described earlier. The second stage will be discussed in Chap. 3.

2.4 Key Steps in the New Technology Needs Assessment

2.4.1 Overview of Steps

The main steps and issues for conducting a TNA are summarised in Fig. 2.3.

Integrating development and climate into climate strategies requires taking a developing country's sustainable development priorities as a starting point and using these priorities as criteria for identifying strategic sectors for climate change mitigation and adaptation and achieving development goals. After the initial organisation of the assessment, these are the second and third step in Fig. 2.3. As argued by CCAP (2010), embedding the action-based processes into such a long-term national framework would increase the coherence of the action portfolio.

In the fourth step in Fig. 2.3, stakeholders are familiarised with technologies within each of these priority sectors or measures for mitigation and adaptation using a range of approaches, including the online platform *ClimateTechWiki*

---

[5] At its sixth meeting on *Development and Transfer of Technologies* (Bonn, Germany, 19–20 November 2010), the EGTT endorsed the updated *Handbook for Conducting Technology Needs Assessment for Climate Change* (UNDP 2010).

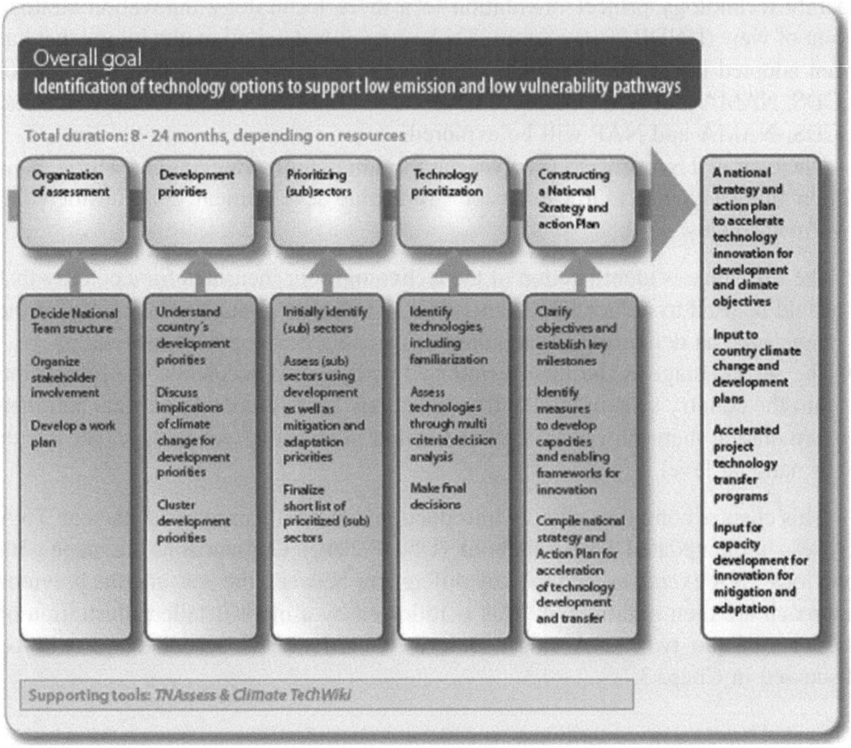

Fig. 2.3 Key steps of the TNA process (UNDP 2010, p. 8)

(see Sect. 2.4.2 and Fig. 2.4). An initial list of technologies is generated and then structured according to whether the technology or measure is available in the short term or the medium to long term and whether it is a small-scale or large-scale technology (these categories are further explained later in this chapter). A multi-criteria decision approach is then used to prioritise a portfolio of technologies and measures in the priority sectors in support of the country's sustainable development.

The prioritisation of technologies or measures within each priority sector is based on a benefit-to-cost ratio over all the sustainable development and climate and/or adaptation benefits. This produces a summary table of portfolios of technologies or measures per priority sector with their sustainable development benefits, costs of roll out and climate/adaptation benefits. This can be used for input to a national strategy and for meta-analysis across countries (see also Chap. 4). These steps are described in more detail later in this chapter.

Within the process, it is important that the voices of different stakeholder groups are heard so that stakeholders' knowledge and concerns are incorporated. Care is taken to have a participatory approach with stakeholders right from the start that supports their 'buy in' in the process, including their role in the eventual implementation of prioritised low-emission and climate-resilient technologies and actions.

Finally, it is important that the approach takes account of the uncertainties that surround the choices. Assessing development and climate change mitigation and adaptation needs implies that decisions are taken for a relatively long period of time, e.g., 20 years, so that stakeholders need to develop a feeling not only for what is happening now but also for what might happen in the future. For instance, a country with a relatively small tourist industry could expect this sector to become bigger in the next two decades and among the larger GHG emitters. Such expectations can be included in the analysis and stakeholders could identify options for making areas suitable for tourism more climate resilient.

The second stage, represented by the fifth step in Fig. 2.3 (described more fully in the next chapter), is to move from the technologies and measures to strategies and action plans. This is done by examination of the existing system for the innovation of the technology or measure followed by identification of barriers and blockages in the system by stakeholder groups. Activities to overcome these problems are then generated by the group, and these form the basis of strategies that are structured according to core elements, such as network creation, policies and measures, organisational and behavioural change, market support activities, education and training, and international co-operation and handling intellectual property rights. The activities can be prioritised and these form the basis of the action plan that provides information on:

- Why an activity is important?
- Who should do it?
- When?
- How it will be monitored and verified?
- How much it will cost?

This process is repeated for other priority technologies. The activities can then be aggregated and rationalised to form strategies and action plans at the (sub)sector or national levels.

2.4.2 What is Different in the 'New' TNA Process?

Before moving to a more detailed explanation of how the new TNA process works, first, it is explained what the differences are with the former TNA process and how the lessons learned from earlier needs assessments between 2002 and 2008 in developing countries (see Appendix A) have been incorporated. Principally, as explained earlier, the updated TNA process (UNDP 2010):

- is firmly based on the sustainable development priorities of the country; and
- Extended from a mainly technology *project* implementation orientation to a more technology *innovation* strategy point of view to accommodate the scale of action envisaged to lower GHG levels or increase climate resilience.

Fig. 2.4 Homepage ClimateTechWiki.org

Another addition in the new TNA process is the specific attention for country stakeholders' *familiarity with technologies* or lack thereof. As was concluded by the EU-funded study ENTTRANS (2008),[6] but also in UNFCCC (2009), there can be gaps in people's awareness of and familiarity with potentially useful technologies (see Box 2.2). Therefore, a new online platform, the *ClimateTechWiki*,[7] was devised that supplements other information sources to aid familiarisation

[6] The study "Promoting Sustainable Energy Technology Transfers: Converting from a Theoretical Concept to Practical Action" (ENTTRANS) was carried out for the European Commission during 2006–2007 by the consortium: Joint Implementation Network (JIN, the Netherlands), Cambio Climático y Desarrollo Consultores (CC&D, Chile), Practical Action (Kenya), Asian Institute of Technology (AIT, Thailand), Tel Aviv University (ICTAF, Israel), Kunming University of Science and Technology (KUST, China), Energy Delta Institute (EDI, the Netherlands), Power Production Company (PPC, Greece), and National Technical University of Athens (NTUA, Greece). ENTTRANS' objective was to analyse how transfer of low-emission energy sector technologies to developing countries could be promoted through the Clean Development Mechanism (CDM). The study was carried out in five countries: Chile, China, Israel, Kenya and Thailand.

[7] ClimateTechWiki is available at http://climatetechwiki.org. Accessed 21 September 2011.

(see Fig. 2.4). The platform contains a range of descriptions of technologies for mitigation and adaptation with practical examples and case studies. It presents additional information on technologies that are either in pilot or pre-commercialisation phases or newly emerging or established. Such information could cover cost data, technology performance (e.g., efficiency, capacity factors, lifetime and degree of technical sophistication required for manufacturing, installation and operation) and whether the technology can be sufficiently relied on.[8]

Box 2.2 Gaps in technology familiarity and awareness

There are two main aspects to lack of familiarity with or limited awareness of technologies, as discussed in ENTTRANS (2008) and UNFCCC (2009). First, some respondents in the ENTTRANS case study countries had never heard of some technologies or did not know anything about specific technologies, such as what it could deliver and whether it was available. This meant that they were not confident in making assessments so that potential technologies would not feature in the final lists.

Furthermore, stakeholders reported examples of how the assessment of technologies was coloured by historic experience. If a new technology had been badly implemented in the country before, for whatever reason, then this created an automatic bias against it for some respondents. Added to this is the fact that people tend to anchor in what they know and are familiar with, which implies that the adoption of new technologies has to overcome this resistance to change in the decision-making process. As Winksel et al. (2006) pointed out: 'Organisations operate in embedded socio-technical networks and tend to re-invest in established competences: disruptive technologies (e.g., renewable energy) rarely make sense to incumbents, so their development tends to be left to small outsider organisations'.

The new TNA process explicitly focuses on both *mitigation* and *adaptation*. By doing so, it is acknowledged that reducing developing countries' vulnerability to climate change is as much a priority as mitigation of climate change. Through adaptation measures, sustainable livelihoods can be ensured and ecosystems on which people depend protected. This will require adaptation measures to increase countries' resilience and for this both market and non-market technologies will be required. Potential areas where adaptation strategies will be necessary are health and social systems, agriculture, biodiversity and ecosystems and production system and physical infrastructure, including the energy grid. Although the process to prioritise measures for adaptation is largely similar to that for mitigation

[8] Other sources for technology familiarisation that could be used are: UNEP guidebooks (http://tech-action.org/guidebooks.htm), IEA Technology Roadmaps (http://www.iea.org/subjectqueries/keyresult.asp?KEYWORD_ID=4156) and study tours, expert lectures and demonstration projects.

technologies (e.g., similar development context), a parallel process has been suggested to take into consideration that both processes may involve different stakeholders and could focus on different areas and sectors. Furthermore, contrary to mitigation technologies, it is not always clear and unambiguous what are technologies for adaptation. The TNA handbook underlines that 'for adaptation the issues of interest tend to impact across … sectors in particular ways. For example, for agriculture projected climate change may mean a water shortage and irrigation problems with implications for the location of agriculture, crop yields and livestock' (UNDP 2010, p. 31).

The updated TNA process uses *multi-criteria decision analysis* (MCDA) to support decision making for the stakeholders as they select the technologies (this is explained in further detail in the next section). For the purpose of the MCDA analysis, an online tool called *TNAssess* has been specially designed, for the sector and subsector prioritisation and for the technology prioritisation.[9]

A major difference compared to the 'old' TNA approach is that in the updated TNA, stakeholders prioritise technologies on the basis of their *benefits-to-cost* ratio. In this case, the *benefits* have been assessed using the sustainable development criteria and thus cover a range of issues not necessarily amenable to monetisation. Therefore, the new TNA maximises the benefits (including mitigation and adaptation benefits) for a given resource. It is thus in line with the need to deploy restricted resources for maximum benefit across a range of goals, including climate change mitigation, adaptation and sustainable development.

In the updated TNA process, technologies identified for the prioritised sectors are *categorised* according to whether they are available in the short term[10] or the medium to long term[11] and whether it is a small-scale or large-scale technology.[12] This categorisation is to allow ease of comparison of technologies as a small-scale technology cannot be easily compared with a large-scale technology, nor one under development with one that is market ready. Why this is important has been illustrated by ENTTRANS (2008) where a tendency among stakeholders could be observed to give lower scores to technologies that would be available only in the longer term as these were considered less important for the

[9] TNAssess can be downloaded from http://climatetechwiki.org.

[10] Short term means that the technology or measure is either very close to market or is already established in other markets or systems ready for diffusion.

[11] Medium to long term are those technologies that may be still at the RD&D stage with 10–15 years to deployment or are at the pilot or pre-commercial stage with up to 5 years for deployment. These longer term technologies may be considered useful over time as they could deliver more benefits than existing technologies and therefore (a) they will receive a demand push, and (b) they will be part of a longer term strategy for innovation.

[12] *Small-scale* technologies are described as those that are applied at the household and/or community level, which could be scaled up into a programme. *Large-scale* technologies are those that are applied on a scale larger than household or community level.

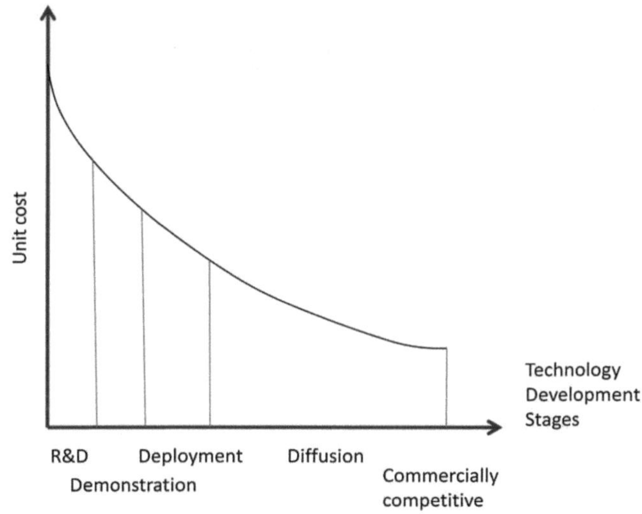

Fig. 2.5 Technology innovation learning curve (EGTT 2009)

time being. Furthermore, stakeholders may, because of their professional background, have a relatively strong focus on, e.g., small-scale technologies only. Categorisation of technologies could thus enable a better comparison of technologies and prevent a biased focus on one particular technology or measures category while overlooking other, also potential beneficial technologies or measures. As a caveat to identifying technologies within categories, it is acknowledged in the updated TNA handbook that the terms short, medium and long term, as well as small and large scale are context specific. For instance, a technology that is commercially viable in one market may not have reached this stage in another market (UNDP 2010, p. 41).

Finally, categorising technologies in terms of short-, medium- and long-term availability is also related to the facility in the new TNA process to formulate strategies for prioritised technologies that are in different development stages if required (this is explained in further detail in the next chapter). Figure 2.5 illustrates this by showing a learning curve for technology innovation:

- Technologies that are still in a process of research, development and demonstration (RD&D) have relatively high-unit costs (longer-term technologies).
- When a prototype has been successfully demonstrated and the technology been successfully deployed in the market, unit costs become lower (medium-term technologies); so that
- The technology could eventually be manufactured and sold commercially competitive (short-term technologies).

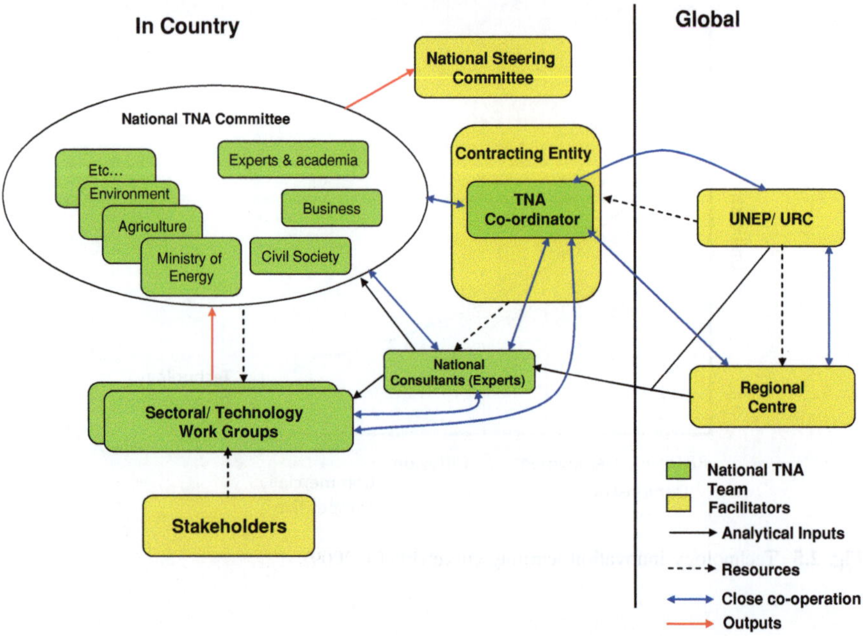

Fig. 2.6 TNA organisation in GEF/UNEP TNA Project (Agbemabiese and Painuly 2011)

2.5 Assessing Technology Needs: How Does It Work?

2.5.1 Organising the Process

A first step in a TNA is the organisation of the process. This includes a decision on who will be responsible for the process, such as a ministry or interministerial committee with experts from all relevant ministries and/or agencies (UNDP 2010; Agbemabiese and Painuly 2011). As an example, Fig. 2.6 shows how the TNA process conducted in the GEF/UNEP TNA Project is organised. It shows how stakeholders from different sectors are represented in sectoral and technology work groups and work on the assessment together with national experts (on, e.g., sectors, technologies and overall country strategies), a national TNA committee, a process co-ordinator and the steering committee. The figure also shows the international and regional support organised under the GEF/UNEP TNA project.

Organising the process in a participatory setting with stakeholders from a broad range of public and private sectors and supporting services has a number of advantages. It can lead to transfer of new, especially local, knowledge and insights on specific technology challenges and opportunities that might otherwise have been missed in a TNA. Furthermore, as mentioned earlier, exposing stakeholders to proposed actions for technology development and transfer in early decision-making stages and letting them actively take part in the process contribute to awareness building and provide some level of 'buy-in' into future technology strategies and action plans. For example, farmers and their communities would make use of technologies to adapt to climatic patterns that have resulted from the technology prioritisation process they have been involved in (UNDP 2010, Chap. 2). Box 2.3 presents an indicative list of possible stakeholders to participate in the TNA process.

Box 2.3 Possible stakeholder groups for TNA

1. Government departments with responsibility for policy formulation and regulation (e.g., power supply) and vulnerable sectors (e.g., agriculture).
2. Private and public sector industries, associations and distributors that are involved in the provision of GHG-emitting services or are vulnerable to climate change impacts.
3. Electric utilities and regulators.
4. Within the private sector, technology users and/or suppliers who could play a key local role in developing/adapting technologies in the country.
5. Organisations involved in the manufacture, import and sale of technologies for mitigation or adaptation.
6. The finance community, which will likely provide the majority of capital required for technology project development and implementation.
7. Households, communities, small businesses and farmers that are or will be using the technologies and who would experience the effects of climate change.
8. Non-Governmental Organisations (NGOs) involved with the promotion of environmental and social objectives.
9. Institutions that provide technical support to both government and industry (e.g., universities, industry RD&D, think tanks and consultants).
10. Labour unions, consumer groups and media.
11. Country divisions of international companies responsible for investments important to climate policy (e.g., agriculture and forestry).
12. International organisations/donors.

Source UNDP 2010, Chap. 2.

2.5.2 Awareness Building and Identifying Development Priorities

As explained earlier in this chapter, in the updated TNA process, a country's development priorities for the short, medium and longer term are used as criteria for selecting strategic sectors for low-emission and climate-resilient development

Fig. 2.7 Stakeholder group in discussion during a TNA (photo courtesy Jorge Rogat)

and for prioritising technologies within these sectors. Such priorities can be derived from several sources, such as already existing strategic documents on climate change measures and sustainable developments (e.g., Sustainable Development Strategies, 5 Year Plans and National Communications) (UNDP 2010, Chap. 3). Identification of these priorities enables country stakeholder groups to revisit existing views of what is important for the country's sustainable development, so that all further decisions in the TNA process can be related to this (Fig. 2.7).

In addition to identifying development priorities, it is also recommended in the updated TNA process that possible climate change impacts on the country are considered. This is obviously important for identifying required measures for climate change adaptation, but a changing climate could also have an impact on selecting priority technologies for mitigation and development goals. For instance, should because of a changing climate hydro resources in a country become smaller, then smaller scale hydro technologies may be more suitable than large-scale hydro power (UNDP 2010, Chap. 3). Given the large uncertainties that surround estimating climate change impacts on a country, a range of possible climate change outcomes may be analysed to develop a feeling for possible climate change impacts to the country and how this would affect technology choices.

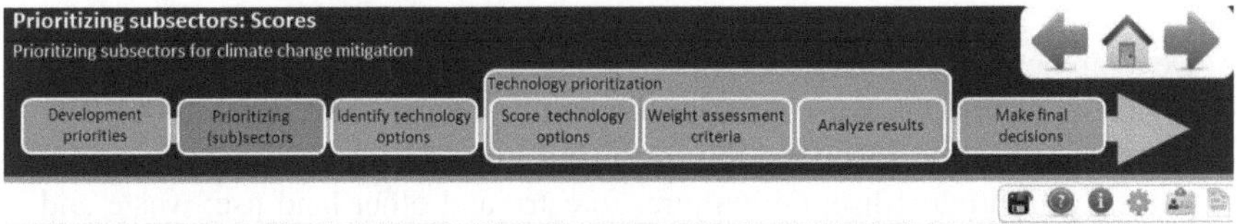

Subsector	Environmental Development Priorities Score		Economic Development Priorities Score		Social Development Priorities Score		GHG Emission Reduction Potential	
Cooking appliances	5 - Extremely desirable	R	3 - Moderately desirable	R	5 - Extremely desirable	R	2 - Fairly desirable	R
Heating and cooling	2 - Fairly desirable	R	3 - Moderately desirable	R	2 - Fairly desirable	R	4 - Very desirable	R
Mainly wet cement production	4 - Very desirable	R	4 - Very desirable	R	0 - No benefit	R	4 - Very desirable	R
Landfills, etc.	4 - Very desirable	R	4 - Very desirable	R	0 - No benefit	R	4 - Very desirable	R
Public transport	2 - Fairly desirable	R	3 - Moderately desirable	R	4 - Very desirable	R	2 - Fairly desirable	R
government buildings	1 - Faintly desirable	R	3 - Moderately desirable	R	1 - Faintly desirable	R	1 - Faintly desirable	R

Fig. 2.8 Prioritisation of (sub) sectors by scoring climate and development contributions (The figure shows an example of a scoring table for climate and development benefits within a (sub) sector. A stakeholder group can consider an improvement in a sector in terms of GHG emission reduction extremely desirable (score 5), in terms of economic development fairly desirable, etc. The highest score in this simple multi criteria decision analysis example is 5 (very desirable) and the lowest score would be 0 (no benefit))

2.5.3 Prioritising Technologies for Strategic Sectors

Once development priorities have been identified with a view to the short, medium and long run, stakeholder groups could identify sectors in the country where improvements (e.g., investments in new low-emission technologies) would result in the strongest combined climate and development benefits (Box 2.4 explains possible ways for sector categorisation). This could be done by first characterising the existing situation in sectors and exploring expected developments for the future. Subsequently, based on these characterisations, stakeholder groups could identify for each sector the potential GHG emission reductions and/or climate change vulnerability improvements, as well as potential economic, social and environmental development benefits. Figure 2.8 illustrates how the benefits could be 'valued' (scored) with help of a simple MCDA, including a rationalisation of values given. These scores could enable stakeholders to take decisions on what are strategic country sectors for low-emission and climate-resilient development.

Box 2.4 Possible sector categorisations for mitigation and adaptation
In a TNA, sectors can be identified by using countries' own sector categorisation, such as, for instance, the sectors identified in the National Communications to the UNFCCC. Countries can also use the classification applied by the UNEP Carbon Finance Group for categorising projects under the Clean

Development Mechanism (CDM).[13] A third option is to follow the *2006 IPCC Guidelines for National Greenhouse Gas Inventories* (IPCC 2006). The latter classification identifies the following main sectors: energy; industrial process and product use; agriculture, forestry and other land use; waste and other sectors.

These main sector categories are further divided into:

- Activities, e.g., 'fuel combustion activities' within 'energy', and 'product uses as substitutes for ozone-depleting substances' within 'Industrial processes and product use'; and
- Sub-sectors, e.g., transport, energy industries.

Possible sectors for adaptation as applied in past TNAs include health and social systems, agriculture and fisheries, coastal zones and water.

Once the priority (sub)sectors have been identified, stakeholders could move on with considering potential technologies (and measures) for mitigation and adaptation for these (sub)sectors. Within the different categories explained earlier,[14] technology options and measures can be identified and information on them gathered. The TNA process subsequently helps country stakeholders to 'personalise' this information to the country's decision context. For instance, stakeholders can estimate the technical potential of a technology option within the country (i.e., if there were no technical and implementation barriers) by concluding that, e.g., based on the country's water resources, hydro power could potentially produce 30% of the country's electricity needs.

The benefit of translating generic technology information to national contexts was also illustrated in the ENTTRANS (2008) study which showed that several stakeholders seemed to assume that technologies that had not been used in their country before would be more expensive than existing technologies and would therefore be more risky. This resulted in lower suitability scores for these technologies. 'Personalisation' would thus help stakeholders to obtain a clear view of the potential role of technology options and measures within each prioritised (sub)sector and related benefits.

At this stage of the assessment, stakeholders can use the information gathered to assess technology benefits, in terms of how they could contribute to the country's development goals, reduce GHG emissions or make the country less vulnerable for a changing climate. Again, just as with the (sub)sector prioritisation, the criteria used for scoring the benefits could be derived from the development priorities determined at the beginning of the process. In addition, financial performance and costs of technology options and measures can be assessed by, e.g., analysing internal rates of return.

---

[13] See http://cdmpipeline.org/cdm-projects-type.htm. Accessed 23 September 2011.

[14] Small scale/short term, small scale/long term, large scale/short term, and large scale/long term.

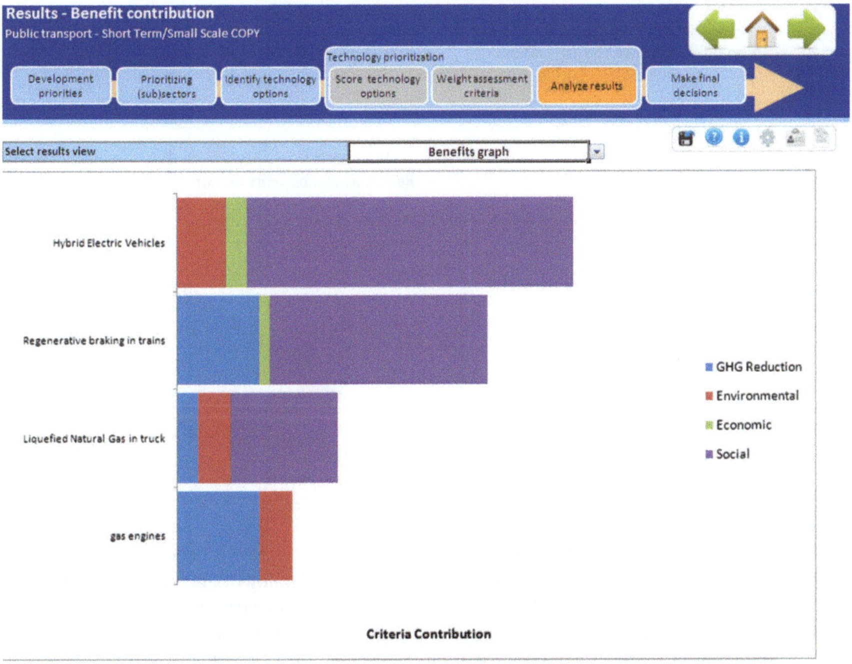

Fig. 2.9 Weighted scoring results for example technology options category (Example taken from the support tool *TNAssess,* available at http://climatetechwiki.org.)

After valuing (scoring) technology benefits, stakeholders can determine the relative importance of the criteria by weighting them. For example, if, for a particular criterion, the difference between the most and least preferred technologies is very small, stakeholders could give a low weight to this criterion, as it would not make much difference which technologies would eventually be chosen. However, if this small difference was thought to be still significant in effect, then a higher weight is given. Similarly when the difference between technology scores for one criterion is large, stakeholders can find this difference very important and therefore assign a large weight to it.[15]

Figure 2.9 shows an example of the possible outcome of the process of scoring the technology options or measures on the benefit criteria and weighting these criteria. The final assessment of an option is the sum, over all the criteria, of the weight times the scores.

---

[15] For instance, if in a country with high unemployment one technology will have strongly positive employment impacts, whereas another technology will have hardly any employment benefit at all, then stakeholders can give a large weight to these scores so that the technology with the strong employment benefit would eventually receive a higher ranking.

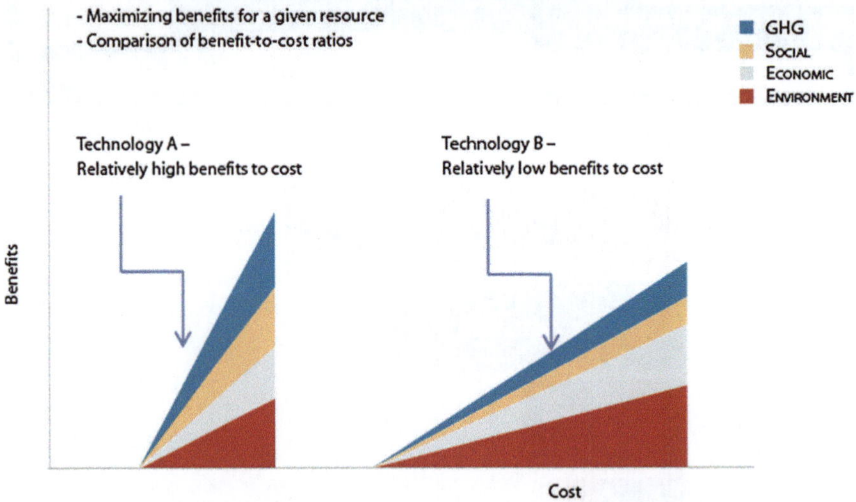

Fig. 2.10 Example of cost-benefit ratio in TNA (This diagram expresses benefits on a scale from 0 to 100 and the costs in monetary values. In this example, technology A is preferred over B as it has relatively low costs and high benefits. Other forms of cost-benefit assessments are also possible and further information about these and why this MCDA was chosen in the updated TNA process, can be found in UNDP (2010, pp. 53–55 and pp. 136–142).)

Not only does the diagram show which technologies within a sector (in this case, short-term and small-scale technologies in the Public transport subsector) have the highest weighted score but it also shows how this score has been built up with climate, environmental, economic and social development benefits.

The robustness of the resulting lists of prioritised technologies per category within a (sub)sector can be checked by sensitivity analysis on different opinions among stakeholders and by exploring uncertainties and the difference they make to the final result. Finally, for making final decisions on priority technology options, benefits from technology options can be compared with a technology's capital and operational costs. An example of such a benefit-to-cost analysis is presented in Fig. 2.10.

The main output of an assessment of low-emission and climate-resilient technology needs for sustainable development, such as in the new TNA process, is a portfolio of technology options and measures that have been assessed as potentially delivering the largest combined climate and sustainable benefits within the country's strategic sectors for low-emission and low-vulnerability development. Table 2.1 shows a hypothetical example of an output table for prioritised technologies for cooking in the subsector of *Residential and offices* in a developing country. The table shows the potential benefits and costs of technology options should they be implemented in the subsector at their technical potential.

Table 2.1 Example of TNA summary table for prioritised cooking technologies in subsector of 'Residential and Offices' in a developing country

Priority technologies identified for cooking in Residential and offices sector	Potential GHG abatement until 2025 at (sub)sector level	Benefits identified from multi criteria decision analysis for technology in TNA	Estimated lifetime costs per technology within (sub)sector (US$)
Short-term/small-scale technologies			
Biogas for cooking and electricity	3.4 Mt CO_2-eq	Improved health because of reduced in-house smoke	17,000,000
		Reduced drudgery for women and children because of reduced need of firewood	
		Reduced poverty at farms	
Charcoal production for cooking and heating	2.7 Mt CO_2-eq	Enhanced carbon sink and moisture reservoir	25,000,000
		Enhanced household energy security	
		Greater entrepreneurial opportunities created through sales of poles and firewood	
		Time spent daily on gathering fuel wood is saved for use in more productive activities	
Short-term/large-scale technologies (none prioritised in this TNA)			
Long-term/small-scale technologies			
Solar cookers	3.8 Mt CO_2-eq	Time savings which results from the reduction in wood gathering	34,000,000
		Build and emphasise links with women's empowerment by creating new organizations led by women	
		The impact of solar stoves on the household economy depends on the organisation of the household economy and the extent to which the household is linked to the wider economic network	
		Improvement of health conditions, promotion on equitable access to energy and poverty alleviation	
Long-term/large-scale technologies (none prioritised)			

Source authors' example based on UNDP (2010, Chap. 5, Tables 5.3–5.6).

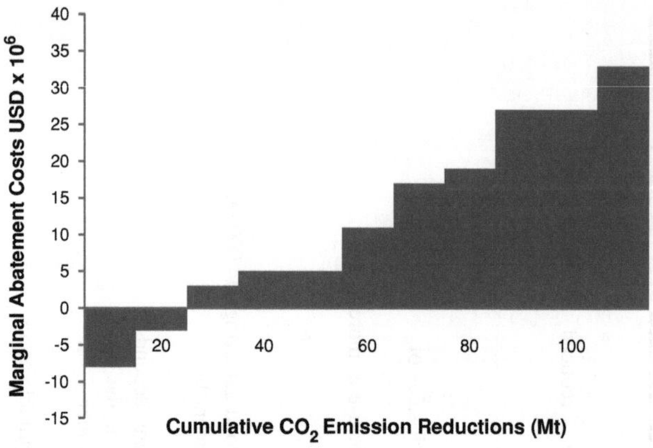

Fig. 2.11 Example of marginal abatement cost curve (*source* authors' example)

2.5.4 Comparison with Cost Curves for Prioritisation

As an alternative approach to selecting suitable technologies with help of an MCDA, marginal abatement cost curves have been used, even though these were originally designed for estimating abatement potentials. In essence the principle is simple. A graph is generated for a country where mitigation measures are costed and put in order of increasing cost for each next (marginal) tonne of GHG reduced (as shown in Fig. 2.11).

Technologies are selected basically in terms of cost efficiency for GHG emission reduction, with the cheapest option first. Some technologies can provide positive monetary savings and this is taken into account both in cost curves and in the MCDA benefit-to-cost approach explained earlier.

In comparison with the MCDA approach as described previously, there could be several problems with using cost curves because they explicitly rate on the cost efficiency for reduction of GHGs with no consideration of sustainable development benefits implying that only cost efficiency of reductions is important. They compare technologies with costs at a specific time which are soon outdated and compare, on the same curve, costs and reductions from, e.g., biogas compared with carbon capture and storage, etc., which are not readily comparable. There can also be large uncertainties in the calculation of reductions. Other problems that could occur in the generation and use of cost curves for technology prioritisation are summarised in Box 2.5.

It is also not clear how the development benefits are taken into account and traded off against this cost efficiency when cost curves are used for a low-emission strategic assessment. The new TNA process, by contrast, maximises over all the benefits for a given cost and is therefore more overall cost efficient for resource allocation.

It has always been recognised that cost curves have advantages and weaknesses and there has recently been some debate about the appropriateness of cost curves for some applications. They have recently been used in the context of Reduced Emission from Deforestation and Degradation schemes (REDD) and for carbon abatement assessment where their use has been strongly criticised for failing to take account of the complexity of many initiatives with negative consequences for deforestation (Greenpeace 2011). This complexity also applies to technologies to be transferred.

Box 2.5 Marginal abatement cost curves and project level assessment[16]
As an example of a Marginal Abatement Cost (MAC) curve, consider the case of energy supply technologies. A cost-effectiveness analysis will essentially develop a GHG abatement cost curve that will rank each technology in the order of its cost-effectiveness for reducing a tonne of CO_2-equivalent emissions. This ranking is typically represented in the form of a curve, as shown in Fig. 2.11.
The identification of priority (sub)sectors and technologies could use as a criterion US$/tonne GHG abatement or MAC curves. A MAC curve would calculate for a country or a group of countries the cost of an additional tonne of GHG emission reduction. These costs depend on the technology with which that marginal emission reduction is achieved. A cost-effectiveness analysis based on costs/tonne GHG abatement could be carried out at the project or plant level, and it would involve total capital costs and operating and management costs divided by the project's total GHG emission reduction. This could be expressed as an annual cost/GHG benefit.
Each point on this curve represents the cost-effectiveness of a given technology relative to the cumulative GHG emission reduction potential achieved when compared with the technology currently used in the country. The points on the curve appear sequentially, from most cost-effective in the *lower left area* of the curve to the least cost-effective options located higher in the cost curve in the *upper right area.*
There are several sources of MACs and it is important to be aware of the following caveats on their use:

1. MAC curves are generated by an analyst. They may not represent the full picture in terms of all abatement technology options to meet development needs. It may not be valid to use them out of the context and time in which they were derived.
2. Many MAC curves cover mainly CO_2 reduction and baselines against which reductions are calculated may be uncertain.[17]

[16] This box has been reproduced from UNDP (2010, pp. 53–55).

[17] The US EPA (US 2006) study covers options for reducing GHGs other than CO_2, including methane. The MAC curve developed by Bakker et al. (2007) combines a large set of bottom-up country abatement studies and covers a large share of abatement options in all sectors and (sub)sectors, including electricity, industry, transport, buildings, waste, agriculture, forestry and land use for most non-Annex I countries. Inevitably, however, these cost curves do not include the full set of mitigation options. The McKinsey and Company (2009) cost curves also have a broad sectoral coverage.

2. In many other studies, mainly the electricity supply (sub)sector is analysed though some industry efficiency, transport, and forestry may also be included (e.g., Bakker et al. 2007). The focus tends to be mainly on large-scale technologies used in centralised grid systems. The decision on which technology is selected is made by the authors who have constructed the MAC curve graph.
3. In some studies, decisions are based on model simulations (Ellerman and Decaux 1998) and expert judgment to derive abatement potential and average costs. Technologies tend to be bundled (e.g., energy efficiency measures) so that individual technologies are not explicitly analysed.
4. Some data on which the MAC curves are based can be quite old. New technologies for low-emission and low-vulnerability development may not be included and studies can become out of date quite quickly.
5. The original data for calculations may cover a range of methods and assumptions that are not necessarily all robust or compatible.
6. In some cases, no-regret options are not identified so that these activities, which would save money and reduce emissions but face other implementation barriers, do not appear on the cost curve. However, not appearing on the MAC curve does not mean these options do not exist. In the case of demand side technologies to reduce GHG emissions from fossil fuel combustion, there are many technologies that have negative costs (i.e., there are net societal benefits from introducing the technology as opposed to net societal costs).
7. Traditionally, the cost calculations used in constructing MAC curves do not take into account co-benefits of mitigation options, e.g., for air quality. In a proper societal abatement cost assessment, these should be included, resulting in significantly lower abatement costs for many options (see Johnson et al. 2009). These are included in the MCDA.
8. Cost curves also compare technologies irrespective of size, e.g., biogas is compared with carbon capture and storage and irrespective of technology innovation stage (e.g., RD&D or diffusion). This is not useful in identifying technologies that currently may have high costs but that deliver multiple benefits of interest to the country and therefore may be subsidised until costs fall through dissemination into the market or system.

2.6 Meeting the Challenge: Transferring Technologies and Measures for Maximum Climate and Sustainable Development Benefits

In this chapter, we have explained the challenge of identifying low-emission and climate-resilient technologies in light of developing countries' short-, medium- to long-term sustainable development objectives.

We have discussed how climate and development policy making has increasingly become interrelated and that climate policy measures in developing countries will only stand a chance of success if in line with development goals. Finally, we have discussed the key role of technology development and transfer in low-emission and climate-resilient development.

The first stage of the TNA process was then described involving the alignment of sustainable development and climate goals for the selection of priority subsectors. The selection of priority technologies in each subsector categorised according to size and stage of development in terms of availability in the short or medium to long term was then described using an MCDA allowing estimation of overall benefits across climate and sustainable development goals. These benefits were then compared with costs and technologies selected on the basis of maximising the benefit-to-cost ratio. The new tools of ClimateTechwiki and TNAssess to support the selection process were described. Where possible some comparisons with other approaches such as cost curves have been made.

The main outputs from this first stage of the TNA process are a portfolio of priority technologies in priority subsectors; an assessment of their overall sustainable development and climate benefits; the costs of the technology at the unit and potential in the subsector, and a comparison of the costs of the technology at the relevant scale to the benefits for maximising the benefit-to-cost ratios for final selection as shown in Table 2.1.

However, we have also underlined, based on earlier experience with TNAs conducted in developing countries, that identifying technologies alone is not the whole story but will need to be supported by overarching strategies for creating an enabling environment for successful deployment and diffusion of the prioritised technologies. At this stage of a needs assessment, country groups can take different perspectives.

For example, the groups could mainly aim at implementing the prioritised technologies as projects. In that case, the main focus would be on identifying technology project barriers and how to address these barriers in the technology project level strategy.

Obviously, when the number of such projects is sufficiently large within a country, it can have an effect on the rest of the economy; such projects could help to improve the overall investment climate in the country for low-emission and low-vulnerability development. However, a project-level perspective runs the risk that projects become stand-alone initiatives and that the diffusion of the technology innovation at the required scale will not occur (ENTTRANS 2008, Chap. 8).

Instead of a project perspective, a country could in their TNAs also take a more strategic perspective by looking at a technology's technical or economic potential and analyse what would be needed to make application of the technology at that scale possible.

The description of the process continues in the next chapter with an analysis of what would be needed to make a priority technology work in the country, either as a project or diffused through a sector or at the national level. For a technology project, barriers would need to be identified and addressed to make a project work, whereas for a sector or country-level strategy, the analysis would focus on improving systems and markets so that priority technologies can be applied at the desired scale.

In the next chapter, how the actions can be identified for addressing technology barriers and bottlenecks and how they can form inputs for formulating technology development and transfer strategies for a technology, within a sector and nationally, will be explained as well as capacity building and finance needs for that.

References

Agbemabiese, L., & Painuly, J. P. (2011). *Technology needs assessments—(Funded by GEFGEF under Poznań Strategic Programme on Technology Transfer)*. UNEP-DTIE and UNEP Risoe Centre. Presentation at UNFCCC workshop on Technology Needs Assessments, Bonn, Germany, 1–2 June 2011. Retrieved June 5, 2011 from http://unfccc.int/ttclear/pdf/Workshops/Bonn%202011/-TNAPresentationUNEPURC-%20UNFCCC-May%20312011.ppt

Anderson, S. (2011). *Climate change and poverty reduction*. Climate & Development Knowledge NetworkNetwork (CDKN). Policy Brief, Retrieved August 9, 2011 from http://cdkn.org/wp-content/uploads/2011/08/CDKN_poverty-reduction_FINAL.pdf

Bakker, S. J. A., Arvanitakis, A. G., Bole, T., Brug, E. van de, Doets, C. E. M., & Gilbert, A. (2007). *Carbon credit supply potential beyond 2012—A bottom-up assessment of mitigation options*. ECN report ECN-E-07-090, Petten ECN. Retrieved September 17, 2011 from http://www.ecn.nl/publications/default.aspx?nr=ECN-E–07-090

CCAP (2010). *Transportation NAMAs: A proposed framework*. The Center for Clean Air Policy, Washington, D.C., USA. Retrieved September 22, 2011 from http://www.ccap.org/docs/resources/924/CCAP_Transport_NAMA.pdf

EGTT (2009). *Future financing options for enhancing the development, deployment diffusion and transfer of technologies under the convention*. FCCC/SB/2009/INF.1. Retrieved September 17, 2011 from http://unfccc.int/resource/docs/2009/sb/eng/inf02.pdf

Ellerman, A. D., & Decaux, A. (1998). *Analysis of post-kyoto CO_2 emissions trading using marginal abatement curves*. MIT Joint Program on the Science and Policy of Global Change Report 40. Retrieved September 23, 2011 from http://web.mit.edu/globalchange/www/MITJPSPGC_Rpt40.pdf

ENTTRANS (2008). *Promoting sustainable energy technology transfers through the CDM: Converting from a theoretical concept to practical action*. European Union Sixth Framework Programme. Retrieved September 23, 2011 from http://jiqweb.org/images/stories/mifiles/downloads/jin/ENTTRANSd2.pdf

Figueres, C. (2011). *Statement by Christiana Figueres at the high-level inter-parliamentary debate on climate change*. Brussels, Belgium, 19 April 2011. Retrieved September 21, 2011 from http://unfccc.int/files/press/statements/application/pdf/110419_speech_inter-parliament_brussels.pdf

Greenpeace (2011). *Bad influence—How McKinsey-inspired plans lead to rainforest destruction.* Retrieved October 7, 2011 from http://www.greenpeace.org.uk/sites/files/gpuk/Greenpeace_BadInfluence_Report_LOWRES.pdf

IEA (2002). *World energy outlook 2002*. International Energy Agency, Paris. Retrieved August 15, 2011 from http://www.iea.org/WEO/docs/weo2002_part1.pdf

IPCC (2006). *2006 IPCC Guidelines for national greenhouse gas inventories*. Task Force on National Greenhouse Gas Inventories. Retrieved September 21, 2011 from http://www.ipcc-nggip.iges.or.jp/public/2006gl/index.html

Johnson, T., Alatorre, C., Romo, Z., & Liu, F. (2009). *Low-carbon development for Mexico*, The World Bank. Retrieved September 23, 2011 from http://siteresources.worldbank.org/INTLAC/Resources/Medec_final_Oct15_2009_Eng.pdf

LDC Expert Group (2009). *National adaptation programmes of action: Overview of preparation, design of implementation strategies and submission of revised project list and profiles.* UNFCCC. Retrieved September 10, 2011 from http://unfccc.int/resource/docs/publications/ldc_tp2009.pdf

Legros, G., Havet, I., Bruce, N., Bonjour, S. (2009). *The energy access situation in developing countries. A review focusing on the least developed countries and Sub-Saharan Africa*. UNDP and WHO, New York. Retrieved September 15, 2011 from http://content.undp.org/go/cms-service/stream/asset/?asset_id=2205620

McKinsey & Company (2009). *Pathways to a low-carbon economy*. Version 2.0 of the Global Greenhouse Gas Abatement Cost Curve. Retrieved September 15, 2011 from https://solutions.mckinsey.com/ClimateDesk/default.aspx

OECD (2009). *Integrating climate change adaptationadaptation into development co-operation: Policy guidance*. ISBN-978-92-64-05476-9. Retrieved September 13, 2011 from http://www.oecd.org/dataoecd/0/9/43652123.pdf

Practical Action (2010). *Poor people's energy outlook 2010*. Rugby, UK. Retrieved August 15, 2011 from http://practicalaction.org/ppeo2010

UNDP (2010). *Handbook for conducting technology needs assessment for climate change*, Retrieved September 23, 2011 from http://unfccc.int/ttclear/pdf/TNATNA%20HANDBOOK%20EN%2020101115.pdf

UNFCCC (1992). *United Nations Framework Convention on Climate Change*, FCCC/INFORMAL/84. Retrieved September 3, 2011 from http://unfccc.int/resource/docs/convkp/conveng.pdf

UNFCCC (2002). *Report of the Conference of the Parties on its seventh session,—Addendum part two: Actions taken by Conference of the Parties*, FCCC/CP/2001/13/Add.1, Decision 4/CP.7. Retrieved September 21, 2011 from http://unfccc.int/resource/docs/cop7/13a01.pdf#page=22

UNFCCC (2008a). *Report of the Conference of the Parties on its thirteenth session*. Held in Bali from 3 to 15 December 2007. FCCC/CP/2007/6/Add.1. Retrieved September 17, 2011 from http://unfccc.int/resource/docs/2007/cop13/eng/06a01.pdf#page=3

UNFCCC (2008b). *Development and transfer of technologies*. Decision 2/CP.14, FCCC/CP/2008/7/Add.1. Retrieved September 21, 2011 from http://unfccc.int/resource/docs/2008/cop14/eng/07a01.pdf#page=3

UNFCCC (2009). *Second synthesis report on technology needs identified by Parties not included in Annex I to the Convention*. FCCC/SBSTA/2009/INF.1. Retrieved June 15, 2011 from http://unfccc.int/resource/docs/2009/sbsta/eng/inf01.pdf

UNFCCC (2011). *Enhancing the implementation of the results of TNAs*. Background Paper II for UNFCCC workshop on technology needs assessments, Bonn, Germany, 1–2 June 2011. Retrieved June 3, 2011 from http://unfccc.int/ttclear/pdf/Workshops/Bonn%202011/BG%20paper%20II-implementing%20TNAs.pdf

US EPA (2006). *Global mitigation of non-CO_2 greenhouse gases*. United States Environmental Protection Agency, Office of Atmospheric Programs (6207 J), Washington, D.C., USA. Retrieved September 23, 2011 from http://www.epa.gov/climatechange/economics/downloads/GlobalMitigationFullReport.pdf

Winskel, M., Mcleod, A., Wallace, R., & Williams, R. (2006). Energy policy and the institutional context: Marine energy innovation systems. *Energy Policy, 33*(5), 365–376.

Chapter 3
Challenge 3: Creating Low-Emission and Climate-Resilient Strategies to Accelerate Innovation for Sustainable Development

Abstract In this chapter, the complexity of the technology transfer and innovation process is discussed with the need to not only deliver technologies but also facilitate successful adoption through supportive activities to overcome barriers in the system. The next challenge, therefore, is to create the low greenhouse gas emission and climate-resilient strategies and action plans to accelerate innovation for sustainable development on the scale and timescale required for achieving climate and development goals. This chapter describes the TNA process that enables identification of the system blockages or market barriers based on characterising the existing system or market. Stakeholders can then generate the activities for overcoming these weaknesses to form a strategy for innovation of a technology. When taken together with other priority technologies in a sector, the identified innovation activities can form a sector strategy, or when scaling up further, form a national strategy. Action plans for implementation provide the basis for the costs, timing, responsibilities and monitoring, verification and reporting as a basis for funding allocation and management.

3.1 Challenge 3: Introduction

Earlier in this book, we have seen that, to avoid serious climate change, there is an urgent need to significantly reduce GHG emissions globally. Despite this, climate policy negotiations under the UNFCCC are slowing, partly because a full global climate coalition with developed and developing countries has not been formed yet. It was concluded in Chap. 1 that this is partly because of the fact that negotiations have for a long time paid insufficient attention to how climate goals could be aligned with countries' domestic development objectives.

W. van der Gaast and K. Begg, *Challenges and Solutions for Climate Change*, Green Energy and Technology, DOI: 10.1007/978-1-84996-399-2_3,

Chapter 2 dealt with how this may be transformed by focussing on integration of sustainable development priorities into climate objectives. Transferring low-emission technologies and measures has been recognised as an important way forward to deliver mitigation and adaptation benefits. By choosing, implementing and diffusing into the country's system low-emission and climate-resilient technologies and measures that also deliver sustainable development benefits, resources could be targeted to maximise both climate and development benefits.

The challenge in this chapter focusses on how to enable technology transfer in a country on the necessary scale for a 'green revolution' in the priority sectors, so that the development and climate benefits can be delivered, including poverty alleviation, in as short a timescale as possible.

Large-scale change needs to occur at all levels, both in terms of measures for changing attitudes and behaviour and in terms of development and technology transfer systems. These changes have to be grounded in a participatory approach and, with regard to equity and sustainable development goals, within the cultural context of the country.

However, large-scale change, either change of behaviour or in terms of technology adoption, normally tends to be a slow process (e.g., requiring 30 years). Part of the challenge is to accelerate this process.

For example, Gruebler (1997) examined the process of technology transfer in terms of the patterns of diffusion of innovation at a macro level. He found that, for a range of technological innovations, they all followed the three stages of invention, diffusion and saturation, as identified earlier by Schumpeter (1939). He also showed that the time between the initial invention/demonstration and adoption varies with how radical the change is in terms of the sociological and organisational and institutional system changes involved. In addition, these innovations needed other critical innovations to be successful, such as, for example, the development of road systems and oil pipelines for supplying the fuel for the adoption of cars.

The challenge, therefore, involves addressing the sociological, institutional and political weaknesses of the transfer and innovation system and their interaction over time. For example, there may be:

- Situations in which existing system players try to block change;
- An inadequate technology base and supply chain;
- A lack of infrastructure;
- Insufficient market or system actors; and
- Conflicting policies and poor networks.

The complexity of the process should also be recognised. The active players in bringing about the change may alter over time, and what was appropriate at the early introductory stages has to change at the later stages of adoption of technologies at a larger scale. Multiple players co-operating in different ways over time will be needed.

In Chap. 2, we suggested that the new Technology Needs Assessment (TNA) process can provide a way forward for integration of development and climate goals. We illustrated this with the approach for prioritising sectors and technologies in line with a country's sustainable development priorities. This provides a portfolio of technologies for transfer into the country system selected by maximising the ratio of climate and development benefits to the technologies' costs.

This chapter deals with the next stage of the new TNA process to address the third challenge of this book:

> To create the low GHG emission and climate-resilient strategies and action plans to accelerate innovation needed to achieve sustainable development and climate goals on the scale and timescale required.

The new TNA provides a participatory process for characterising the existing system and its blockages to devise a set of activities for overcoming these weaknesses. These activities should in effect accelerate the innovation and thus can form a strategy for innovation of a technology. When taken together with other priority technologies in a sector, the identified innovation activities can form a sector strategy, or when scaling up further, form a national strategy.

Action plans for implementation of the strategies can then be devised at the relevant scales to enable successful accelerated transfers into the country system. The TNA process, therefore, has been designed to try to provide an in-depth understanding of the complex multi-dimensional, multi-actor system for enabling and supporting transfers within a country context.

Some early examples of how this process could work, including initial market maps to characterise existing markets or systems, identification of barriers and accelerating actions, are given from the ENTTRANS (2008) study. The chapter concludes with a comparison with other approaches being devised under the UNFCCC as far as information is available.

First, the complexity of technology transfer is discussed in the next section to provide the substantive basis for the new TNA process.

3.2 Technology Development and Transfer

3.2.1 Introduction

Innovation involves the development and introduction of a new technology or measure into a country system and its gradual adoption by society. Innovation into country systems has been recognised as a key driver of economic development in all countries for many years.

Europe's track record, however, is of being good at research but not innovation. The European Institute of Innovation and Technology (the EIT, launched in 2008) is designed to improve this by encouraging the creation of new businesses and education in entrepreneurship. Internationally, the World Bank Group's *info*Dev initiative (Information for Development Program) is in the process of

> "designing and launching a network of Climate Innovation Centers (CICs) to accelerate the development, deployment, and transfer of locally relevant climate technologies. The CICs should increase developing country Green Growth and job creation by financing and supporting the competitive and profitable involvement of cleantech Small and Medium sized Enterprises (SMEs) in local and international supply-chains and markets".[1]

In 2011, InfoDev announced a new initiative in Africa.

As shown in Chap. 2, technologies prioritised for climate change mitigation and adaptation in a country could be in different innovation stages: research, development and demonstration (RD&D), deployment and diffusion. With the help of a learning curve, these stages were characterised by showing that technologies that are still in a process of RD&D have relatively high unit costs, whereas unit costs become lower as the technology diffuses through the market and eventually becomes commercially competitive and saturates the market.

The innovation process can be hit and miss with some promising technologies failing to make it past the development stage. How this could happen is, among other sources, explained by the concept of Valley of Death developed by the World Bank (UNFCCC 2011). According to this concept, technologies could fail after the development stage because of perceived risks, unfamiliarity with the technology, inefficient enabling environment, etc. As a result, the usual replacement of public sector support (during the RD&D stage) with private sector involvement (for deployment and diffusion) may not take place, so that the technology development could be stopped in a 'valley of death' (see also Chap. 5).

Earlier in this book we explained how the role of technology transfer for climate change mitigation and adaption has been defined in Art. 4.5 of the United Nations Framework Convention on Climate Change (UNFCCC 1992). In the context of the article, as described in Chap. 2, further technology transfer support for developing countries has been worked on, such as the Technology Development Framework of 2001 with TNAs (UNFCCC 2002) and the Poznan Strategic Programme on Technology Transfer of 2008 (UNFCCC 2008a). An important benefit of this support is that it helps developing countries avoid being locked into old and less sustainable technologies, by transfer of knowledge, technology and finance. The new Technology Mechanism will now be the focus of technology transfer support for the large-scale changes required.

We also pointed out in Chap. 1, that the term technology transfer is not limited to the transfer of hardware, but also refers to transfer of knowledge and experience, and involves the process of 'learning to understand, utilise and replicate the technology including the ability to decide which technology to transfer and adapt it to local conditions and integrate it with indigenous technologies' (IPCC 2000). However, transfer of knowledge and hardware is not enough for a complete transfer of a technology or measure.

The process involves people with their cultural norms and habits, country systems and markets. Therefore, in addition to the technology base in terms of

---

[1] http://go.worldbank.org/QUSDA36Q51. Accessed 26 September 2011.

manufacturing capacity, supply chain capacity, end-of-life/waste disposal structure, support services and sustainability of the whole process, for successful transfer, attention must be paid, among others, to the host country's social capital, including institutional capacity, skills, networks, norms and educational base.

For the technology or measure to deliver benefits to the communities within the host developing country, it is also imperative that local organisations and communities are involved in the technology implementation process. Therefore, the process of technology transfer has to be tailored to the host country and participatory; technology transfer is not a one-size-fits-all process.

3.2.2 Context for TNA System Approach for Innovation Strategy Generation

3.2.2.1 Brief Review of Literature

There are several literature sets (e.g., innovation, development, socio-psychology, sociology, economics, etc.) that are relevant to innovation. Innovation theories cover a wide range from *Actor-Network* 'theory' (Latour 2005; Law 1992; Law and Hassard 1999) to National Systems of Innovation (NSI) (Lundvall 1985; Lundvall et al. 2002). The theories available can take account of both the system aspects and the social and technical interaction aspects of innovation processes. Furthermore, they consider the perspectives of the end users and their behavioural change. Foxon et al. (2008) take a multidisciplinary approach to understanding innovation using the disciplinary perspectives of the economic, institutional and management literatures. In this review, we do not attempt to cover the full range of literature, but, to place the TNA process in its context, we try to cover some of the main aspects of innovation systems and some of the socio-technical aspects.

Successful innovation depends on a large variety of important aspects. NSI theory focuses on the role of social capital that operates through multi stakeholder networks to perform the information and expertise exchange and collaboration necessary for development of the system (Freeman 1987; Lundvall 1992; Lundvall et al. 2002). However, the particular country context determines the innovation system. Gu and Lundvall (2006) suggest conditions for innovation within a country that include:

- Market regulations and planning;
- Informal rules for collaboration and communication to create new knowledge;
- Robust and diversified systems;
- Good support in terms of institutions and structures (e.g., legal); and
- Learning systems to build competence within the system and within society as a whole.

Other approaches include the technological (Carlsson and Stankiewicz 1991), sectoral (Breschi and Malerba 1997) and regional systems of innovation within a country.

The theory, therefore, supports the importance of the country context on technology transfer. How firms and organisations in different countries interact depends on the differences between countries in terms of language, cultural and business habits. Johnson (1992) pointed out that the cultural rules and habits are important for forming these interrelationships between players. Countries will also have varied regulatory and legal frameworks affecting the systems and markets.

The diversity of stakeholders in the technology transfer process is wide and can include project developers, technology owners, technology suppliers, product buyers, recipients, users of the technology, financiers and donors, governments, international organisations, non-governmental organisations (NGOs) and community groups (IPCC 2000). In addition to these groups, though not mentioned in IPCC (2000), we could mention that research organisations, trade organisations and educational institutions are also important in supporting the transfer process.

The socio-psychological literature considers the behavioural change on the part of the receiver for the process of adoption of an innovation. People must actively adopt the technology and adapt to its requirements. The efficiency of the innovation process, therefore, will also depend on people's experience, values and perceptions, as well as on the economic, social and political environment. Rogers (1971, 1983) suggested that the process of diffusion of a product in a market will depend on the individual buyer characteristics with a range from innovators and early adopters to late majority and, finally, laggards.

Social acceptance plays an increasing role in the market development of an innovation as can be seen in the 'NIMBY' syndrome.[2] It has been proposed that initial broad positive acceptance backed by policies to encourage new adoption and investment and involvement in local planning decisions are all important (Wustenhagen et al. 2007).

3.2.2.2 Policy Intervention for 'System Failure'

Introducing a new product in an existing system or market could create the problem that it is likely to displace an existing product and that the existing players lose income and power. These players will react by blocking the new product as far as possible and interfere with the co-operation and communication aspect of innovation development, or as Gu and Lundvall (2006) stated 'interactive learning possibilities can be blocked and competences destroyed for political reasons' and 'counteractive policies are needed as is stability in the financial and macroeconomic fiscal and monetary policies'. Intellectual property is often used as a tool for blocking new developments. This is called 'system failure' when some external policy interventions are required to address problems blocking the system.

In energy system studies, there is also a need for policy to support a technology push and feed-in market pull for effective system building. Demand-driven

---

[2] NIMBY stands for not in my backyard.

innovation and initially protected niche markets have been proposed to address these problems (e.g., Jovanovic and Rob 1987; Van den Ende and Dolfsma 2002). Policy interventions to make firms consider alternative energy technologies may also be required (Rip and Kemp 1998).

Developing a new technology requires a long period of learning and network building and policy support to maintain a range of different designs. According to Bergek and Jacobsson (2002), creating and maintaining this Design Variety require an inclusive style of system building that aligns well with the Multi Polar World model (see Box 3.1) and the 'open innovation' model (Chesbrough et al. 2006) for the initial research and development stage of innovation.

Box 3.1 Further developments: Multi polar world of innovation

The IPCC (2000) only considers the traditional model of technology transfer where a developed country will develop the technology and widely adopt it before it is transferred to a developing country, i.e., North–South technology transfers. However, this neglects the fact that some developing countries have climate friendly technologies that could be transferred from South to South and also from South to North (Brewer 2007).

A number of barriers can present themselves when using the traditional model of North–South transfers. In the new TNA handbook (UNDP 2010), several key issues are identified:

- The process will be too slow for the urgency of climate change action;
- The technologies have not been designed for developing country conditions; and
- The potential for innovation is being ignored within these countries.

To promote within-country development and innovation into the system, a new paradigm, where the developing countries will be involved as technology developers as well as technology users, is suggested so that South–North and South–South transfers can be made.

This new multi polar world of innovation would address most of the existing problems, enabling developing countries to undertake new technology development designed for their own conditions. Many developing countries already have the capacity, including an educated and skilled workforce, low-cost manufacturing capabilities and demand-driven potential (UNDP 2010, Annex 10).

In this new arrangement, developing countries would participate in the innovation process creating a networked process of commercialisation across developing and industrialised countries allowing research and development contributions to flow from one region to another and back until a new product is achieved. Examples of this are the biofuels industry in Brazil and Chinese experience with the manufacturing of solar and wind power.

Source UNDP 2010.

Based on a review of the innovation and policy literature, as well as stakeholder input, Foxon et al. (2005) and Foxon and Pearson (2007) provided guidance for policy making to support a Systems Innovation (SI) regime with five guiding principles:

1. Stimulate the development and SI policy regime that brings together appropriate strands of current innovation and energy and environmental policy and regulatory regimes;
2. Apply systems thinking and practice engaging with the complexity and systemic interactions of innovation systems and policy-making processes;
3. Advance the procedural and institutional basis for the delivery of SI policy;
4. Develop an integrated mix of policy processes, measures and instruments that cohere to promote SI; and
5. Incorporate policy learning as an integral part of the SI policy process.

This guidance is developed within a UK context, but could serve as guidance for coherent policy making in any country as it supports the range of activities required for acceleration and deals with the complexity of the system.

This brief review illustrates some of the complexity of a technology transfer process and forms the basis for the process proposed in the new TNA for moving from the identification of prioritised technologies for climate and development goals (as in Chap. 2) to strategies and action plans for acceleration of innovation (this chapter).

Further developments in innovation relevant to this are discussed in Box 3.2.

Box 3.2 Delivery Models for technology transfer

The concept of delivery models for energy access has recently been used in the Practical Action (2010) report based on studies under the project *Policy Innovation Systems for Clean Energy Security* (PISCES), funded by the UK's Department for International Development (DfID). This concept describes who may be involved and what may occur in terms of the type of financing and implementation at the project level to overcome barriers to transferring a new technology.

Practical Action (2010) points out that with some new technologies, it is appropriate that development organisations are involved at the early stages and it is through a combination of a range of actors over time that commercialisation may occur. The report lists existing possible delivery models that range from co-operative utility models to decentralised virtual utilities and small enterprise management models. However, though they may be private or community based, they range in terms of the actors and their roles during implementation.

Practical Action (2010), therefore, propose a framework for delivery where they identify the main components for delivery of energy access. The basic model consists of identifying the user, the energy service, the end user energy vector, e.g., gas, and the delivery system. It then guides the user through the main approaches compatible with the information in the basic model relating to the equipment purchase, the ownership model, maintenance model and financing model. It also takes account of the resource rights and the appliances that use the energy with attention to the leadership and management model for the initiative. This latter will tend to involve a number of actors varying over time.

The IPCC (2000) report also recognises that the pathways for transfer will depend on the country context, sector and type of technology. Possible pathways identified in IPCC (2000) are:

- Government assistance programmes;
- Direct purchases;
- Licensing;
- Foreign direct investment;
- Joint ventures;
- Co-operative research arrangements;
- Co-production agreements;
- Education and training; and
- Government direct investment.

In practice, if transfers are to occur on a large scale, then combinations of these types of ventures may be required. These could be mobilised perhaps in combination with international funding under the Technology Mechanism (see Chaps. 4 and 5). Development NGO involvement, especially for energy access for poverty alleviation, will also be essential.

The first step is to explore the existing system for technology transfer and innovation, which will be the focus of the next section.

3.3 Designing Strategies and Action Plans for Accelerated Innovation at Technology, Sector and National Levels

From the preceding discussion, the full transfer and adoption of even one technology into a subsector will entail a long-term strategy for low-emission and climate-resilient sustainable development. Successful implementation of the strategy requires an action plan at the national, sector or technology level, which supports the acceleration of the technology's successful introduction and adoption in the country. The next stage of the new TNA process addresses this challenge as it describes a process to help countries to identify technology acceleration activities for inclusion in a strategy and to prepare action plans for that. This section explains this TNA stage in some detail. The first step in the description is to characterise the existing system for transfer and innovation of a technology.

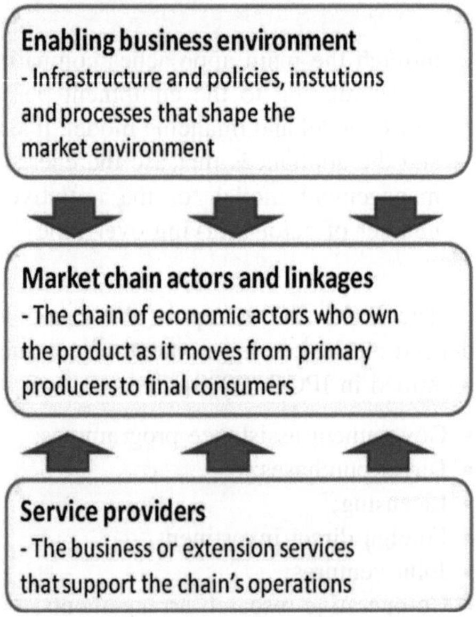

Fig. 3.1 The market map method (Albu and Griffith 2005)

3.3.1 Exploring Existing National Systems: Market Mapping

Before we can identify activities for improving the environment for enabling successful transfer of a technology or measure, we need to closely examine the existing market (e.g., if a technology can be rolled out as a commodity) or system (e.g., if it is a set of measures to adapt to climate change) in which it will have to operate. This allows key players in the market or system to be identified, as well as gaps and system blockages.

This market or system analysis is preferably done in a participatory process with active market players, as well as other appropriate stakeholders, so that the best available knowledge can be generated in the analysis. Furthermore, it supports engagement of the key players and engenders their commitment to the implementation of the programme or strategy eventually developed by them. One way of doing this, which is advocated in the new TNA handbook (UNDP 2010, Annex 13) and applied in the ENTTRANS (2008) study, is to use market or system mapping. It is worthwhile going into a little detail on this and some early examples are provided to illustrate the method and see the power of the approach.

Albu and Griffith (2005) describe the process of mapping by dividing the market map into three core elements: the business enabling environment, the market chain and the market supporting services. These three elements are illustrated in Fig. 3.1.

The *market chain* contains all the main economic actors such as primary producer, importer, traders, processors, input suppliers, energy companies, wholesalers, retailers and customers. The *supporting services* are the business and extension service providers supporting the market chain. The purpose is to identify the needs for services and who the users are. This gives insights on what can be done in terms of supporting services to make the market more efficient. Such services can include financial services, quality control, technical expertise, market information services, etc.

The *enabling business environment* includes the critical factors and trends shaping the market and the operating conditions, such as infrastructure, policies and institutions. The purpose is to identify the trends affecting the business environment and to identify who has the power in the market and who is driving change. In turn, this can then provide information on whom to lobby and help determine opportunities for action. According to Albu and Griffith (2005), the enabling environment encompasses the following:

- Relating to market demand;
- Consumption trends;
- Tax/subsidies and tariff regimes;
- Relating to transformation activities, i.e., costs of doing business;
- Infrastructure constraints;
- Investment policies;
- Transport policies and licensing;
- Technological development;
- Trade regime (import/export);
- Systems of finance;
- Gender roles in business and finance;
- Registration of land and property;
- Legal requirements for contracts;
- Commercial law;
- Business licences and regulation; and
- Standards quality control and enforcement.

The process is conducted through a participatory market chain analysis that results in the creation of a *market map*. In addition, mapping a market also contributes to building trust, negotiating and networking. An example of what a final market map could look like is given in Fig. 3.2.

As Fig. 3.2 shows, the map of the system can be quite detailed and have quantitative information attached to the flows of goods or services or revenues shown by the thin blue arrows. How the supporting system links to the market chain is shown as red arrows.

3.3.2 Mapping for Technology Transfer: Case Study Kenya

The relevance of market mapping for the technology transfer process is that the technique can be used within a participatory environment with stakeholder groups to explore any system. It can identify gaps and blockages within the existing systems in the country for roll-out of technologies or measures for mitigation and

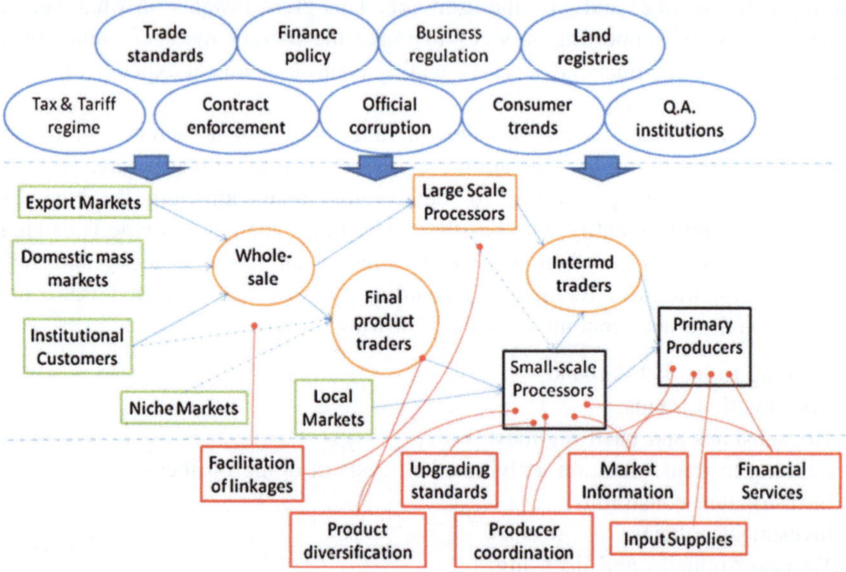

Fig. 3.2 An example market map (Albu and Griffiths 2005)

adaptation. Furthermore, with the technique, a picture can be obtained of what these systems would need to look like for enabling an acceleration of technology development and transfer.

As a part of the ENTTRANS (2008) study for the European Commission (see also Chap. 2), a series of workshops were organised during 2007 to explore existing market systems in Chile, China, Israel, Kenya and Thailand. Before the workshops, technologies had been identified in these countries as suitable for contributing to their prioritised energy services. An example from the ENTTRANS (2008) project is described in the following section to illustrate the technique.

For the workshop in Kenya, the market mapping technique was applied for two prioritised technologies: large-scale concentrating solar power for supply to an electricity grid (CSP),[3] and small-scale biomass gasification stoves for household cooking.[4] The participants were split into two groups, one for each technology.

[3] Concentrating solar power (CSP) systems concentrate the energy from the sun for electricity production. This is done by heating a fluid which is then used to raise steam for a conventional turbine for on- and off-grid electricity provision. These systems can also provide heat, either at high temperatures directly for chemical reactions, e.g., chemical processing, or as a by-product for desalination plants or cooling systems, depending on requirements.

[4] Biomass gasification household stoves work by a high-temperature conversion of biomass in a restricted oxygen environment to a mixture of nitrogen, carbon monoxide, hydrogen and methane. The hydrogen and methane are then burned without emitting pollutants. The technology can be applied in households, institutions (such as schools) and industries where it is used for boiler heating.

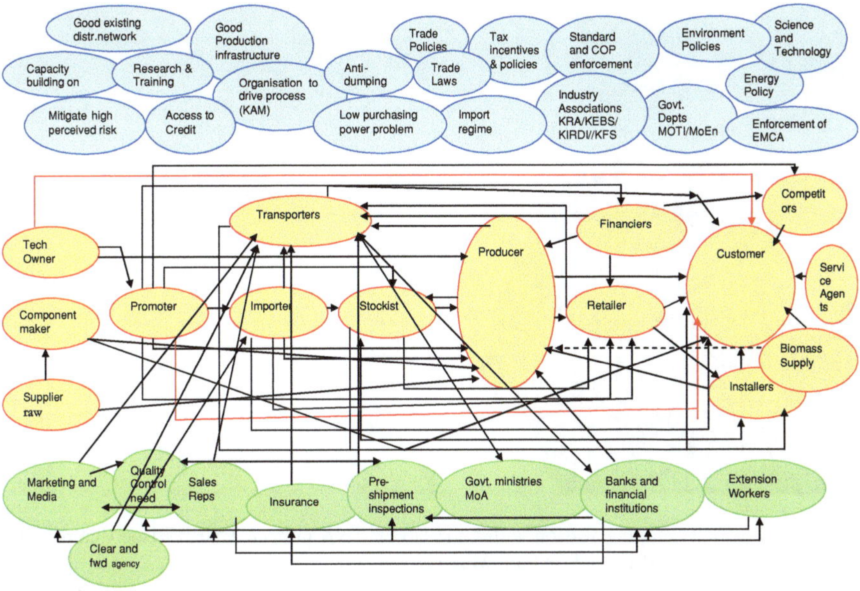

Fig. 3.3 Kenya Small-Scale Biomass gasification Stoves market map (ENTTRANS 2008)

The initial map for the small-scale biomass gasification stoves is illustrated in Fig. 3.3.

The blue ellipses at the top in Fig. 3.3 represent the enabling environment discussions, the yellow ellipses in the centre represent the market chain and the green ellipses in the lower part of the diagram show the supporting services identified by participants. The lines between ellipses show the linkages for the supporting services with the market chain and the interactions between the players in the market chain. More in depth discussions would normally follow this initial map with quantification where possible. This is then used as a basis for discussion on what is working or not working within the system, what is missing and how it might be developed in the future. Figure 3.4 illustrates this process in action.

Taken together, the insights from the market maps in the ENTTRANS (2008) case study countries showed that the enabling business environment and the supporting services along with the links can provide an initial picture of the complexity of the system. Blockages and improvements can be identified to enable full technology transfer.

Through the market mapping exercise, it has also become clear that the maps have elements in common across technologies and countries. For example, the enabling business environment common themes found in ENTTRANS (2008) were as follows:

- Government department co-ordination and red tape;
- Fiscal and legal regulations;

Fig. 3.4 A market mapping session within the UNEP TNA project (UNEP Risoe Centre 2011, photo courtesy of Dr. M'hamed Sedrati, Morocco)

- Trade and research institutions;
- Transparency and trade laws, especially regarding independent power producers;
- For small-scale technologies: local expertise, research, training, financing and risk offsetting;
- For large-scale technologies: tariffs, privatisation and international incentives;
- Policy environment; and
- Public awareness and information.

As a next step following the initial mapping, in the workshops, stakeholders were asked to identify opportunities arising from the market mapping analysis. The key opportunities seen by the participants in the ENTTRANS workshops in utilising the 'new' technologies discussed for the market map included the following trade and sustainability benefits:

- Trade opportunities;
- Economic development and superior technology;
- Electricity load balancing and security of supply;
- Poverty alleviation;
- Increased funding and new policy directions; and
- Improved skills.

Finally, the mapping exercises showed that the country conditions are important along with the type of technology and its size. For example, small-scale technologies tend to need more market players in the chain. ENTTRANS (2008) concluded that for adoption of a small-scale technology by the market, there is a need for either an interface to an existing network for reaching customers or the creation of such a network and interface to it.

3.3.3 Common Blockages Identified in the ENTTRANS Case Study Workshops

Generally, the process of technology transfer has to overcome several barriers (Ellis and Kamel 2007; IPCC 2000). As we are aiming for more large-scale transitions based on the prioritised technologies and measures, this could imply that significant effort would be needed to overcome the barriers in the system.

The market mapping approach applied in the ENTTRANS (2008) case studies has shown that in different countries, there exist a mixture of inefficiencies related to the market chain, the enabling business environment and the supporting service requirements. Table 3.1 lists some common technology transfer inefficiencies identified in the study.[5] Common barriers across technologies and countries include elements such as government department co-ordination and red tape, fiscal and legal regulations, trade and research institutions, business environment and support services, etc. The evidence points to the importance of country conditions along with the type of technology and its size. These insights are a first step in elaborating the complexity and key activities for successful transfer.

In this discussion, technologies in the energy sector have been mainly used to illustrate the process. However, as mentioned already earlier, the market mapping approach can be applied to a wider range of areas, including technologies and measures for mitigation and adaptation prioritised in a TNA. In the next section, we explain how the insights from market maps can be used in a TNA to generate technology strategies and action plans.

3.4 Generating the Strategy and Action Plan

There can be many ways for a country to derive a strategy for technology innovation, but few have attempted a change on the scale required now for avoiding dangerous climate change effects. As explained in Chap. 1, the Cancun Agreements (UNFCCC 2010) mention nationally appropriate mitigation actions (NAMAs) and national adaptation plans (NAPs), along with low-emission development strategies (LEDS). However, details on how these can be formulated have not been worked out yet under the Convention. The following discussion

[5] A more detailed table with other literature sources are given in the TNA handbook (UNDP 2010). ENTTRANS (2008) also identified barriers that are country specific and technology specific.

Table 3.1 Common blockages to technology transfer identified by ENTTRANS (2008)

Enabling environment blockages	
Import procedures need to be simplified and incentivised for these new technologies	Poor infrastructure especially for communication for small-scale project support
Lack of integration across government; for instance, fiscal policies such as tax regimes also need to be aligned to encourage adoption of low-emission technologies	Weak policies and legal framework and enforcement
Standards and enforcement are key	For large-scale electricity supply, concern is over tariffs and feed in tariffs, government incentives and net metering
	Lack of incentives
Market chain blockages	
Lack of technology transfer network	Availability of cheaper high-carbon alternatives
Lack of awareness of stakeholders and for large projects particularly linkages and contacts to external producers	Need to demonstrate unfamiliar technologies and adapt to local conditions
Cost of new technologies and no accounting for externalities	Lack of competition especially in electricity supply
Support services blockages	
Lack of spare parts	R&D support needs to be reviewed
Poor extension services	Finance availability for new technologies and small-scale technologies and measures to offset the additional risks associated with these new technologies
Social and cultural barriers	Gender participation and integration/equity
Lack of media interest	
Monitoring and evaluation	
Capacity building for training manpower and manufacturing	

outlines the bottom up approach taken in the new TNA process (UNDP 2010) for generating these outputs and then looks at the other approaches for developing strategies to address technology innovation for sustainable development goals.

3.4.1 Strategies and Action Plans in the New TNA Process

Once the full map of a market or system is available and the blockages and barriers are identified, it can then be used as the basis for stakeholder discussions to generate the list of activities for addressing these barriers and bottlenecks. These activities will accelerate innovation and therefore can form inputs for technology development and transfer strategies at a range of levels: technology, sector and national level. In this process, account has to be taken of the technology itself, any modifications required for the technology to work under country conditions, the particular skills base needed and supply chains.

In addition, as described in Sect. 3.2, whether the technology is applied on a small or large scale needs to be taken into account, as well as the stage of innovation of the technology: RD&D, deployment and diffusion. Technologies that are still in the process of RD&D need a different enabling environment to technologies that are close to commercial application.

For example, for technologies in the RD&D stage, it is important that research capacity is available within the country with a strong supporting role by public sector developers, and that intellectual property right issues are taken into account. On the other hand, for technologies in the process of deployment into a market, the focus may be on what skills base for operation and maintenance needs to be fostered or how quality control and enforcement for the technologies can be created.

3.4.1.1 Structuring the Accelerating Activities into a Strategy

To manage the list of activities for acceleration, these activities are first of all structured in terms of *core elements*. In the new TNA process, the following core elements are used:

- Measures for creation of stakeholder networks to enable the exchange of ideas and information to ensure dissemination of innovation;
- Policies and measures to promote technology transfers;
- Organisational/behavioural change;
- Market, system support and financial services;
- Skills training and education; and
- International co-operation.

Most attention on innovation or technology transfer in TNA reports in the past focussed on implementing single projects and overcoming the barriers to their success (UNFCCC 2011). The updated TNA process allows both a project technology strategy and a much larger scale technology innovation, such as for a sector or the country as a whole. A technology strategy is generated by identifying activities for accelerating the development and transfer of the technology within the (sub)sector at a desired scale. For example, for improving the enabling environment for technologies for adaptation within the agriculture sector, it could be suggested that the government organises a country-wide training programme for, e.g., crop-rotating techniques. This could increase the knowledge level within the sector of operating the prioritised technology or measure. This activity could then be categorised under the core element 'skills training and education' as an input for a strategy.

This can be repeated for other priority technologies identified for the (sub)sector. Larger scale strategies can be generated by aggregating across technology-level strategies for a (sub)sector level strategy and by aggregating across (sub)sectors for a national strategy.

Accelerating activities in the technology-level strategies can be both technology-specific and be at a more general, sector or national level. Aggregating these activities within their core elements across technologies or across (sub)sectors, as described earlier, is likely to lead to overlaps or redefinitions for similar activities.

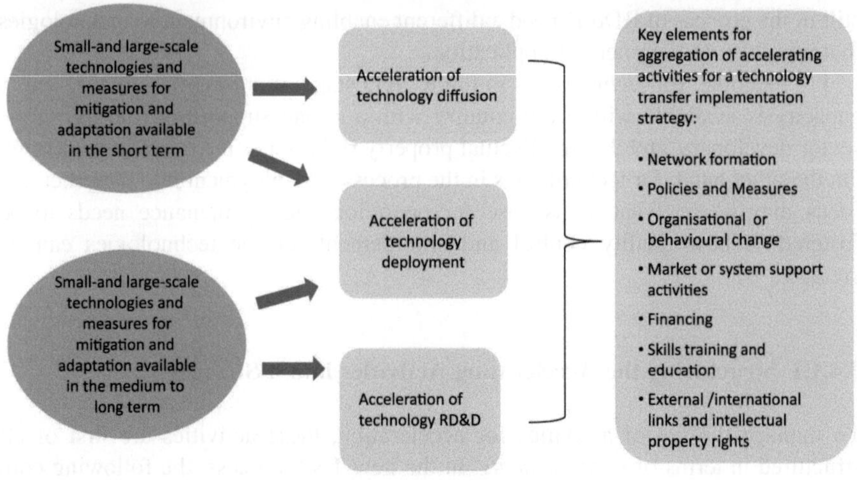

Fig. 3.5 Aggregation into strategies across core elements

Therefore, rationalisation is required when aggregating activities. Depending on the degree of aggregation, this allows for a strategy for successfully implementing technology projects, or formulating (sub)sector-level technology programmes or a national technology innovation strategy. This process is illustrated in Fig. 3.5.

In the process of strategy formulation, technology acceleration measures are identified for market ('hard') and non-market ('soft') technologies in different innovation stages so that a further strategy differentiation can be made for technologies under development (RD&D), ready for deployment and ready for diffusion into the country system at (sub)sector and national level depending on country requirements.

3.4.1.2 Formulating Action Plans

All the accelerating activities in a strategy will be needed eventually, but stakeholder groups can prioritise the accelerating actions for inclusion in an *action plan* for implementing the strategy. The prioritised actions in the action plan are then characterised according to the following questions:

- Why is the action important?
- How can the action successfully be carried out?
- What timeframe is envisaged for the action?
- Who should be responsible for the action under this core element?
- What are required measuring, reporting and verification (MRV) activities for successful action?

Sector: Agriculture							
Specific Technology and category: Crop rotation system – small and large scale – short term							
Innovation Stage: Deployment – Diffusion							
Measure (grouped under core elements)	Priority	Why is it important?	Who should do it?	How should they do it?	Time scale	Monitoring, reporting and verification for measure	Estimated costs
Formation of networks							
Identification of existing networks	1						
Creation of hubs	2						
Policies and measures							
Demand driven innovation policies	1						
Other core elements as listed, e.g., skills training, etc.							
Measure 1 etc	3						

Fig. 3.6 Characterising activities for an action plan (UNDP 2010, p. 85)

- How much would the action cost?

It will be essential to incorporate feedback in the management of the action plan to promote learning by doing and flexibility to respond to a range of contexts. Figure 3.6 illustrates how an action plan can be assembled.

At this stage, information has been gathered about what actions are needed for acceleration of technology development and transfer; how these actions can be incorporated into a technology, sector level and/or national strategy; and how the characterised actions form action plans for implementation of the strategy.

The action plans allow several important outputs to be developed:

- The costs of each of the supporting accelerating activities for the innovation strategy are identified. Thus, the overall process generates costs both for these innovation support activities and the costs of the technology or measure as in Table 2.1 in Chap. 2.
- On this basis, resource allocations can then be made and justified.
- The strategies and action plans incorporate timing of the actions so that plans can also be developed over time depending on priority actions now compared with more long-term (e.g., education) actions.

- The measuring, reporting and verification (MRV) actions are worked out in the action plan as is the accountability for the actions being monitored and managed. These outputs can also be presented as the basis for funding applications (see Chap. 5).

These supporting activities in the strategy are not add-ons to the deployment of the technology but are intrinsic to success as they allow the technology or measure to overcome barriers to accessing the market or system.

The trust thus built and the relationships formed during the mapping process helps the group to continue working as a functioning network during an action phase. This last phase has to be taken up by the country stakeholders as an ongoing activity.

The description of the process given here is by necessity brief and a more detailed explanation is given in the TNA Handbook (UNDP 2010, Chap. 6). To bring about change on the scale required and make real progress, the process will require much more effort and time than in earlier simpler project TNA exercises, where the focus was more on technology-level analysis (as described earlier in this chapter).

3.4.2 Comparing TNA with Low Emission Development Strategies

As explained in Chap. 1, the core objective of a low-emission development strategy or plans (LEDS)[6] is to integrate economic development with climate policy objectives (Clapp et al. 2010; Ellis, Baker and Lemma 2009). Since the introduction of the concept in the Bali Plan of Action in 2007 (UNFCCC 2008b), several initiatives have been taken to give shape to this emerging concept. Clapp et al. (2010) present an overview of studies in both developed and developing countries showing organisational aspects of a LEDS in terms of ownership of the process within the country, data requirements and possible synergies with other processes such as national planning reports, National Communications, TNAs, etc.

In the Cancun Agreements, LEDS is mentioned under 'Enhanced action on mitigation', albeit differently for developed and developing countries (UNFCCC 2010). Whereas the COP has decided that developed countries *should* develop LEDS, developing countries *are encouraged* to develop LEDS in the context of sustainable development. Similar to a TNA, a LEDS, therefore, remains a voluntary exercise for developing countries, and is broadly seen as an enabling exercise that could, for instance, help prioritise NAMAs and support their implementation. Therefore, LEDS can support formulating portfolios of NAMAs, but is not seen as a precondition for decisions on NAMAs in a developing country.

[6] As explained in Chap. 1, the terms low-carbon and low-emission development strategies are used interchangeably in literature sources. In this book, we use LEDS.

Although several LEDS studies have been carried out in a number of developing countries, there is no clearly established process yet for this concept. The focus of studies has, therefore, been on gaining insights into the technical, institutional and policy aspects of LEDS, e.g., how to harness synergies with other national policies and strategies, how LEDS could develop into an overarching sustainable development country strategy, the involvement of high-level policy makers, avoidance of overlaps between stakeholders' responsibilities, requirements for MRV, etc.

Based on a review of studies, Clapp et al. (2010) identify the following key steps for LEDS:

1. Formulation of emission reduction and development goals by assessing the current situation of a country and aligning development with climate goals;
2. Establishing an institutional framework with inter-ministerial participation and consultation of stakeholders;
3. Prioritisation of policies by identifying barriers to low-carbon development, considering policy synergies and trade-offs and exploration of policy interactions across sectors;
4. Identification of sources of finance and co-ordination of funding disbursement for implementation of the prioritised policies; and
5. Tracking progress towards goals and feed lessons learned into updates of LEDS.

In comparison, the new TNA process has the following key steps, as seen in this and the former chapter:

1. Formulation or clarification of sustainable development goals in the short, medium and long term;
2. Formation of TNA government structure and implementation team with stakeholder engagement strategy;
3. Identification of priority (sub)sectors for the sustainable development goals that also provide important mitigation or adaptation benefits;
4. Identification of priority technologies within the priority (sub)sectors for meeting the goals;
5. Characterisation of existing system and identification of gaps and blockages in the existing market or system into which the technology or measure will be developed, deployed or diffused;
6. Identification of accelerating activities for successful transfer of the technology/ies;
7. Structuring and aggregation of accelerating activities at the technology level for a project implementation strategy, across technologies in a (sub)sector for a (sub)sector innovation strategy and across (sub)sectors for a national innovation strategy; and
8. Construction of action plans to implement the strategies including who should do it, when, how, how much it will cost, MRV requirements and management with feedback.

In terms of overall objectives, there are several overlaps between the new TNA and LEDS. In both processes, the analysis is grounded in a country's vision on the short- to long-term development priorities, and these priorities are used as criteria for further assessment. Furthermore, both processes aim at providing strategic input for formulation of national policies for sustainable development in the country, including MRV provisions.

Whereas a new TNA focuses explicitly on prioritising low-emission or adaptation technologies and measures (such as behavioural change) within strategic (sub)sectors for achieving a country's long-term policy objectives, in LEDS the technology focus is more implicit. LEDS studies appear to take a more top-down approach so that mitigation technologies may be selected with the help of marginal abatement cost curves focussing only on technology costs (as discussed in Chap. 2).

Therefore, there is a risk that benefits of a technology and country-specific aspects could be missed and alignment with sustainable development priorities is more difficult. More information is required before a definitive comparison can be made, but from current studies, it is not fully clear whether LEDS take a system view, as the TNA process does, though policies are generated from barriers identified. The LEDS output is the generation of enabling policies and measures that are a key component of success. However, as we explained earlier in this chapter, policies and measures are only a part of the strategic requirements for successful full-scale innovation that will involve activities by many system actors.

Hence, there seems to exist potential for harmonisation of TNA and LEDS process steps. For example, TNA can incorporate policy analysis from LEDS under the policy and measures core element (Sect. 3.4.1) for the formulation of policies for low-emission climate-resilient development.

The new TNA process does not specifically try to locate sources of finance as in the fourth step of an LEDS (see overview from Clapp et al. 2010 in the preceding) for policy implementation, but it does provide information on financial needs and sources. Rather, the new TNA tries to ensure that implementing the accelerating activities in an action plan will allow resource allocation to be conducted, so that finance can be targeted for maximum effect. Figure 3.6 illustrates that each activity in the action plan has to be prioritised and costed. This allows funding agencies a clear view of what would be expected to happen, who is responsible for this and what MRV action will be undertaken. Tracking and feedback are implicit in the management of the action plans.

In such a harmonised TNA/LEDS approach, the stakeholders involved in both processes could be similar, although it could be argued that stakeholders in a TNA could have a wider background within a sector, whereas the stakeholders involved in an LEDS would have a more financial and policy-level background.

How TNA and LEDS, either separately or harmonised, could contribute to formulating NAMAs and NAPs will be discussed, among other topics, in the next chapter.

3.5 Meeting the Challenge: Strategies and Action Plans to Accelerate Innovation for Climate and Sustainable Development Goals

In this chapter, we have proposed a way of meeting the challenge of creating low-emission or adaptation projects, programmes and strategies, with action plans, for accelerating innovation on the scale needed to avoid dangerous climate change. Given the urgency of climate change action and the interrelation between climate and development, we suggest that this has to be carried out within a sustainable development context paying attention to the social, environmental as well as economic aspects and with equity to avoid risks of failure and rejection.

The basis for the new TNA process for innovation strategy development in terms of innovation theory has been elaborated as well as the process described for mapping, with stakeholder groups, the existing system for a technology transfer. From this, the groups could identify blockages, gaps and opportunities. This has been illustrated with an example from Kenya. Based on these barriers, the activities to accelerate innovation are identified through discussion. These accelerating activities can be structured into core elements (policies and measures, networks, etc.), aggregated and rationalised to form strategies at the technology level, across technologies at the (sub)sector level and across (sub)sectors for the national level. Strategies can also be formed for the different innovation stages according to the country needs. Action Plans can be prepared for the innovation strategies that will allow costs to be calculated and combined with technology costs for resource allocation for accelerating innovation as well as resource allocation for the technology hardware. MRV is also put in place in the action plans that also allow accountability for the actions. Strategies and action plans over time enable short- and long-term plans to be made.

The new TNA approach was then compared with the LEDS proposals under the climate negotiations as a possible way of elaborating these concepts. However, because of the lack of current information, this could only be an indicative exercise but some merging of the two concepts would seem beneficial.

In the next chapter, we explore further the challenge of moving forward the provisions in the Cancun Agreements for mitigation and adaptation (TNA, LEDS, NAMA and NAP) under the international climate policy regime to achieve the sustainable development and climate goals through low-emission transfers at the scale and in the timescale required.

References

Albu, M., & Griffith, A. (2005). *Mapping the market: A framework for rural enterprise development policy and practice*. Practical Action report. Retrieved September 27, 2011 from http://practicalaction.org/docs/ia2/mapping_the_market.pdf

Bergek, A., & Jacobsson, S. (2002). The emergence of a growth industry: A comparative analysis of the German, Dutch and Swedish wind turbine industries. In S. Metcalfe & U. Canter (Eds.), *Change, transformation and development: Schumpeterian perspectives* (pp. 197–228). Heidelberg: Springer.

Breschi, S., & Malerba, F. (1997). Sectoral innovation systems. In C. Edquist (Ed.), *Systems of innovation: Technologies*. London: Institutions and Organisations, Pinter Publishers.

Brewer, T. L. (2007). *International climate change transfer: New paradigm and new policy agenda agenda*. Retrieved September 27, 2011 from http://www.mendeley.com/research/climate-change-technology-transfer-new-paradigm-policy-agenda-3/

Carlsson, B., & Stankiewicz, R. (1991). On the nature, function and composition of technological systems. *Journal of Evolutionary Economics, 1*(2), 93–118.

Chesbrough, H., Vanhaverbeke, W., & West, J. (2006). *Open innovation: Researching a new paradigm*. UK: Oxford University Press.

Clapp, C., Briner, G., & Karousakis, K. (2010). *Low-emission development strategies (LEDS): Technical, institutional and policy lessons*. Paris: OECD, IEA. Retrieved September 23, 2011 from http://www.oecd.org/dataoecd/32/58/46553489.pdf

Ellis, K., Baker, B., & Lemma, A. (2009). *Policies for low carbon growth*. ODI: ODI Research Report.

Ellis, J., & Kamel, S. (2007). *Overcoming barriers to Clean Development Mechanism projects*. OECD, UNEP Risoe Centre. COM/ENV/EPOC/IEA/SLT(2007)3. Retrieved September 27, 2011 from http://www.oecd.org/dataoecd/51/14/38684304.pdf

ENTTRANS (2008). *Promoting sustainable energy technology transfers through the CDM: Converting from a theoretical concept to practical action*. European Union Sixth Framework Programme. Retrieved September 23, 2011 from http://jiqweb.org/images/stories/mifiles/downloads/jin/ENTTRANSd2.pdf

Ernst, D., & Lundvall, B-A. (1997). *Information technology in the learning economy: Challenges for developing countries*. DRUID working paper 97–11, Aalborg, Denmark: Department of Business studies.

Foxon, T. J., Pearson, P., Makuch, Z., & Mata, M. (2005). *Policy drivers and barriers for sustainable innovation*. ICEPT Monograph 2005/1.

Foxon, T. J., & Pearson, P. (2007). Overcoming barriers to innovation and diffusion of cleaner technologies: Some features of a sustainable innovation policy regime. *Journal of Cleaner Production*, 16(1), pp. 148–161.

Foxon, T. J., Köhler, J., & Oughton, C. (Eds.). (2008). *Innovation for a low carbon economy: Economic institutional and management approaches*. London: Edward Elgar.

Freeman, C. (1987). *Technology policy and economic performance: Lessons from Japan*. London: Pinter.

Gruebler, A. (1997). *Time for a change: On the patterns of diffusion and innovation*. International Institute for Applied Systems Analysis. Research report RR-97-3.

Gu, S., & Lundvall, B.-A. (2006). China's innovation system and the move toward harmonious growth and endogenous innovation. *Innovation, Management, Policy & Practice, 8*(1–2), 1–26.

IPCC (2000). *Methodological and technological issues in technology transfer—A special report of IPCC working group III*. Intergovernmental Panel on Climate Change Intergovernmental Panel on Climate Change. Geneva, Switzerland. Retrieved September 21, 2011 from http://www.ipcc.ch/pdf/special-reports/spm/srtt-en.pdf

Johnson, B. (1992). *Institutional learning in 'National Innovation Systems': Towards a theory of innovation and interactive learning*. London: Pinter.

Jovanovic, B., & Rob, R. (1987). Demand-driven innovation and spatial competition over time. *Review of Economic Studies, 54*(1), 63–72.

Latour, B. (2005). *Reassembling the social: An introduction to actor-network-theory*. Oxford: Oxford University Press.

Law, J. (1992). *Notes on the theory of the actor network: Ordering, strategy, and heterogeneity*. Retrieved September 28, 2011 from http://www.lancs.ac.uk/fass/sociology/papers/law-notes-on-ant.pdf

Law, J., & Hassard, J. (Eds.). (1999). *Actor network theory and after*. Oxford: Blackwell and the Sociological Review.

Lundvall, B.-A. (1985). *Product innovation and user-producer interaction*. Aalborg: Aalborg University Press.

Lundvall, B.-A. (1992). Introduction. In B. Lundvall (Ed.), *National systems of innovation: Towards a theory of innovation and interactive learning*. London: Pinter.

Lundvall, B. A., Johnson, B., Andersen, E. S., & Dalum, B. (2002). National systems of production, innovation and competence building. *Research Policy, 31*, 213–231.

Practical Action (2010). *Poor people's energy outlook 2010*. Rugby, UK. Retrieved August 15, 2011 from http://practicalaction.org/ppeo2010

Rip, A., & Kemp, R. (1998). Towards a theory of socio technical change. In S. Rayner & E. L. Majone (Eds.), *Human choice and climate change*. Ohio: Batelle Press.

Rogers, E. M. (1971). *Diffusion of innovations* (2nd ed.). New York: Macmillan Publishing Co.

Rogers, E. M. (1983). *Diffusion of Innovations* (3rd ed.). New York, NY: The Free Press.

Schumpeter, J. A. (1939). *Business cycles, a theoretical, historical and statistical analysis of the capitalist process, Vols I and II*. New York: Mcgraw-Hill.

UNDP (2010). *Handbook for conducting technology needs assessment for climate change*, Retrieved September 23, 2011 from http://unfccc.int/ttclear/pdf/TNATNA%20HANDBOOK%20EN%2020101115.pdf

UNEP Risoe Centre (2011). *Technology needs assessment newsletter*, Vol.1, Retrieved June 13, 2011 from http://tech-action.org/TNATNA_NewsJun2011.htm

UNFCCC (1992). *United Nations Framework Convention on Climate Change*, FCCC/INFORMAL/84. Retrieved May 23, 2011 from http://unfccc.int/resource/docs/convkp/conveng.pdf

UNFCCC (2002). *Report of the Conference of the Parties on its seventh session,—addendum part two: Actions taken by Conference of the Parties*, FCCC/CP/2001/13/Add.1, Decision 4/CP.7. Retrieved September 21, 2011 from http://unfccc.int/resource/docs/cop7/13a01.pdf#page=22

UNFCCC (2008a). *Development and transfer of technologies*. Decision 2/CP.14, FCCC/CP/2008/7/Add.1. Retrieved September 21, 2011 from http://unfccc.int/resource/docs/2008/cop14/eng/07a01.pdf#page=3

UNFCCC (2008b). *Report of the Conference of the Parties on its thirteenth session*, held in Bali from 3 to 15 December 2007. FCCC/CP/2007/6/Add.1. Retrieved July 17, 2011 from http://unfccc.int/resource/docs/2007/cop13/eng/06a01.pdf#page=3

UNFCCC (2010). *The Cancun Agreements: Outcome of the work of the Ad Hoc Working Group on Long-term Cooperative Action under the Convention*. Decision 1/CP.16. FCCC/CP/2010/7/Add.1. Retrieved July 17, 2011 from http://unfccc.int/resource/docs/2010/cop16/eng/07a01.pdf#page=2

UNFCCC (2011). *Enhancing the implementation of the results of TNAs*. Background Paper II for UNFCCC workshop on technology needs assessments, Bonn, Germany, 1–2 June 2011. Retrieved June 3, 2011 from http://unfccc.int/ttclear/pdf/Workshops/Bonn%202011/BG%20paper%20II-implementing%20TNAs.pdf

Van den Ende, J., & Dolfsma, W. (2002). *Technology push, demand pull and the shaping of technological paradigms-patterns in the development of computing technology*. ERIM Report Series Research in Management. ERS-2002-93-ORG. Rotterdam, the Netherlands. Retrieved September 26, 2011 from http://www2.eur.nl/WebDOC/doc/erim/erimrs20021022132314.pdf

Wustenhagen, R., Wolsink, M., & Buerer, M. J. (2007). Social acceptance of renewable energy innovation: An introduction to the concept. *Energy Policy, 35*, 2683–2691.

Chapter 4
Challenge 4: Realising the Promise—Rationalising the Current Directions in International Climate Policy Making

Abstract Recent developments in the negotiations have shown an increasing focus on developing strategies for meeting climate goals in the context of countries' sustainable development. Provisions included for that in the Copenhagen Accords and Cancun Agreements are low-emission development strategies (LEDS), nationally appropriate mitigation actions (NAMAs) and national adaptation plans (NAPs). In addition, the updated TNA process supports formulation of strategies with action plans. This chapter discusses possible interlinkages between these provisions as well as with the several climate policy pillars under the Convention, such as Technology Mechanism, Cancun Adaptation Framework, Capacity-Building Framework and Financial Mechanism. The chapter addresses the challenge of how to rationalise these current directions in climate policy making. It provides recommendations for the efficient operation of these pillars by suggesting areas where activities can be integrated so that duplication and wastage of resources can be avoided.

4.1 Challenge 4: Introduction

We started this book with a discussion of the urgency and scale of action needed to address climate change. With the recent development of the *Cancun Agreements* (UNFCCC 2010), the negotiations have moved in the direction of a future climate policy regime where countries could be supported to achieve their economic, social and environmental objectives with low-emissions and increased climate resilience.

The climate policy negotiations have resulted in provisions and mechanisms to help countries identify mitigation and adaptation actions within the context of sustainable development. We identified this as our first challenge in Chap. 1.

W. van der Gaast and K. Begg, *Challenges and Solutions for Climate Change*, Green Energy and Technology, DOI: 10.1007/978-1-84996-399-2_4,

We then discussed the challenges that a country faces in terms of prioritising actions (such as technologies, techniques or behavioural change) for reaching both sustainable development and mitigation and adaptation goals. Using this as a basis, we then explored the challenge of how to formulate strategies and action plans to accelerate innovation of these actions on the desired scale within a country.

The next challenge, which will be discussed in this chapter, is how:

To rationalise the current directions in international climate policy making to provide coherent and efficient support to developing countries in devising and implementing strategies and action plans for low-emission technology transfers to deliver climate and sustainable development goals.

The chapter discusses in further detail the provisions for mitigation and adaptation in the Cancan Agreements, as well as mechanisms introduced for technology transfer, capacity building and finance. This is followed by a discussion on how these provisions and mechanisms might interact with the technology prioritisation and strategy formulation processes discussed in the previous chapters. Finally, based on the insights thus gained, we explore ways forward under an international climate policy regime for low-emission and climate-resilient development. In the following sections, each provision or mechanism is taken in turn and explored in more detail.

4.2 Supporting and Delivering the Cancun Agreements

4.2.1 Overview of New Provisions and Mechanisms

As explained in Chap. 1, according to the Cancun Agreements, 'developing country Parties will take *nationally appropriate mitigation actions (NAMAs)* in the context of sustainable development, supported and enabled by technology, financing and capacity-building, aimed at achieving a deviation in emissions relative to "business as usual" emissions in 2020' (UNFCCC 2010, para 48).

Furthermore, all Parties are invited to enhance action on adaptation under the *Cancun Adaptation Framework*, through 'planning, prioritising and implementing adaptation actions, including projects and programmes (NAPs), and actions identified in national and subnational adaptation plans and strategies, national adaptation programmes of action of the [least developed states], national communications, technology needs assessments and other relevant national planning documents' (UNFCCC 2010, para 14b).

A new Adaptation Committee will promote adaptation actions under the Convention (UNFCCC 2010, para 20).

In addition, developing countries are encouraged to develop *low-emission development strategies or plans (LEDS)* in the context of sustainable development (UNFCCC 2010, para 65).

Another major development is the formation of the *Technology Mechanism* from the former expert group on technology transfer (EGTT). The Technology Mechanism has been established to facilitate 'enhanced actions on technology development and transfer to support action on mitigation and adaptation to achieve the full implementation of the Convention' (UNFCCC 2010, paras 113–115 and 117). The objective of the Technology Mechanism is to become the centrepiece of an agreed outcome on technology transfer and development under the Convention (UNFCCC 2010, para 117).

In summary, the new mechanisms that have started to emerge from the UN Climate Conferences (COPs) in Copenhagen and Cancun (see also Chap. 1) include the Cancun Adaptation Framework, the Capacity-Building Framework and the Technology Mechanism. The new provisions for identifying actions for mitigation and adaptation include the NAMAs and NAPs, and LEDS in addition to the existing new process of conducting technology needs assessment for climate change (TNA, as explained in Chaps. 2 and 3).

The Cancun Agreements have not specified how to formulate NAMAs and an LEDS, and for formulation of NAPs have referred to developing countries' 'experience in preparing and implementing national adaptation programmes of action' (UNFCCC 2010, para 15).[1] For the Technology Mechanism, some basic structure was agreed in the Cancun Agreements (as explained in Sect. 4.3), but it is also still being developed as is the Cancun Adaptation Framework and Capacity-Building Framework. Therefore, in the following discussions, how the new climate policy provisions and mechanisms may be developed to fulfil their promise is explored along with how the new TNA process could support their development and operation.

4.2.2 Delivering NAMAs and NAPs

The current absence of an established process to formulate NAMAs under the Convention (UNFCCC 2010) has resulted in a considerable diversity among communicated NAMAs (ClimateFocus 2011). For instance, an investment in a low-emission technology in a project or programme could be considered a NAMA, as well as actions undertaken to improve the functioning of markets or systems for successful development and transfer of low-emission technologies. What will be considered NAMAs is to be decided by the developing country concerned. The only main requirement for NAMAs as determined by the Cancun Agreement is that they need to be formulated *in the context of sustainable development* (UNFCCC 2010, para 48).

[1] According to UNFCCC definition, national adaptation programmes of action (NAPAs) provide a process for Least Developed Countries to identify priority activities that respond to their urgent and immediate needs to adapt to climate change—those for which further delay would increase vulnerability and/or costs at a later stage. http://unfccc.int/national_reports/napa/items/2719.php. Accessed 28 September 2011.

Adaptation has always been a key element of the UNFCCC and has been considered through the *Buenos Aires Programme of Work on Adaptation and Response Measures* (UNFCCC 2004) and the *Nairobi Work Plan* (UNFCCC 2007) to improve understanding and assessment of impacts and make informed decisions on practical adaptation actions and measures. The *Bali Action Plan* (UNFCCC 2008) laid the foundation for further development by identifying adaptation as a key building block for strengthening future response to climate change.[2]

To help developing countries to identify and take measures for adaptation to climatic change, the Cancun Agreements contain a decision on the establishment of the *Cancun Adaptation Framework* as explained earlier. It was agreed that 'enhanced action on adaptation is urgently required to enable and support the implementation of adaptation actions aimed at reducing vulnerability and building resilience in development country Parties, taking into account the urgent and immediate needs of those developing countries that are particularly vulnerable' (UNFCCC 2010, paras 11–35).

To address this urgency, the Cancun Agreements recommend the following for enhancing actions on adaptation (UNFCCC 2010, Chap. 2):

1. Supporting the implementation of NAPs with prioritised actions for adaptation. This support consists of exchanging information on existing channels for support provided and received by countries for adaptation actions, support to least developed countries (LDCs) to formulate and implement NAPs and a 2-year programme to consider approaches to address loss and damage associated with climate change impacts in developing countries that are particularly vulnerable to the adverse effects of climate change. NAPs prioritise urgent and immediate adaptation needs for LDCs.
2. A request to developed countries to provide developing countries with long-term scaled-up predictable new and additional finance, technology and capacity building to implement adaptation actions, plans, programmes and projects at local, national, sub-regional and regional levels.
3. Organisation of support at the global level (Adaptation Committee), regional level (e.g., establishing regional centres and networks) and national level (establishing and/or designating national-level institutional arrangements).
4. NAPs need to be undertaken with a view to integrating adaptation into relevant social, economic and environmental policies and actions.
5. Vulnerable groups as well as relevant multilateral, international, regional and national organisations, the public and private sectors, civil society and other relevant stakeholders need to be involved when undertaking and supporting enhanced action on adaptation at all levels.

[2] Potential areas for adaptation are: protection of coastal zones; health improvement; infrastructure improvement; terrestrial ecosystems; water resource protection and management; and agriculture, livestock and fisheries.

A TNA focuses on both measures for mitigation and adaptation using a participatory process with involvement of, e.g., vulnerable groups. Based on the discussions in Chaps. 2 and 3, we would suggest that the updated TNA process could be a suitable way forward for the formulation of NAMAs and NAPs. We show this by the following:

- Technology needs are identified in the light of a country's sustainable development objectives as well as climate change imperatives.
- Strategies are identified for country context specific[3] systems of technology development and transfer at the technology, subsector and national levels.
- These strategies incorporate activities on capacity-building and finance needs, policies and measures, networks, organisational change, supporting activities for the system, and intellectual property rights requirements for successful technology development and transfer.
- Action plans can be developed for implementing the strategies that allow time planning, allocation of responsibilities and resources, and measuring, reporting and verification (MRV) to maximise the benefits.

Putting together a portfolio of NAMAs and formulating NAPs could, therefore, benefit from work undertaken by TNAs with identified actions for technology acceleration.

The new TNA process facilitates resource allocations so that resources not only for the technology equipment or measures but also for its innovation into the country system are accounted for. This can be particularly useful for the efficient use of resources for adaptation from current sources that include the Global Environment Facility (GEF) under the national communications,[4] the LDC fund, the Adaptation fund in relation to the Clean Development Mechanism (CDM) and the Special Climate Change fund (see for these initiatives also next chapter).

A particular aspect of a TNA, which could enhance successful implementation of NAMAs and NAPs, is the identification of MRV requirements in the action plans. This ensures that the actions identified for acceleration of technology development and transfer (e.g., solving barriers and bottlenecks) at the technology level, within the sector and/or nationally are actually implemented and managed.

MRV is needed, for example, to determine the effectiveness of identified actions, make adjustments based on feedback and to assess whether financing has been put to good use (CCAP 2009; Jung et al. 2011). While the Cancun negotiations provided a schedule for establishing guidelines for MRV (particularly for NAMA formulation), these guidelines need to be further developed which is likely to take several years (Jung et al. 2011).

---

[3] Each country has specific national institutional structures and social networks of actors (e.g., technology providers and private project developers). They operate under their respective policies and regulations.

[4] See, for further information about national communications to the UNFCCC, http://unfccc.int/national_reports/items/1408.php. Accessed 28 September 2011.

Whatever level is appropriate for the country, a 'package' of MRV actions could result to ensure successful transfer of prioritised technologies through improvement of the enabling environment for adoption of technologies for mitigation and adaptation. Through this MRV package, TNAs can enhance the implementation of national climate policy actions (e.g., NAMA and LEDS). Box 4.1 shows some examples of how MRV actions identified in a TNA could be used for NAMA (and NAP) MRV.

Box 4.1 MRV actions identified in a TNA and suitable for NAMAs and NAPs

The following possible MRV requirements, which could be identified by a TNA, could form input for formulating MRV guidelines for NAMAs and NAPs:

- Monitor the establishment of stakeholder networks within sectors to ensure dissemination of innovation, report on their meetings, participants, etc., and verify that the networks operate well.
- Monitor the incentives recommended in the strategy for making the system for technologies more efficient (e.g., introduction of feed in tariff), report on the effectiveness of the measures and verify this with recommendation on whether adjustments are needed.
- Monitor the support actions to enhance market support services (advice, financial intermediation, etc.) and verify the impact on technology project investments and programme development.
- Monitor how international links are established to arrange intellectual property rights for particular technologies and verify the reported progress, etc.

4.2.3 Delivering an LEDS

In the Cancun Agreements, LEDS was introduced under 'Enhanced action on mitigation' (UNFCCC 2010, Chap. 3). It is a voluntary exercise for developing countries as is a TNA, and it is broadly seen as an enabling exercise that could for instance help prioritise NAMAs and support their implementation (see also Chaps. 1 and 3). Therefore, as discussed in Chap. 3, while LEDS can support formulating portfolios of NAMAs, it is not seen as a precondition for decisions on NAMAs in a developing country.

The process involved in producing an LEDS was discussed in detail in Chap. 3 and compared with the TNA process that also delivers strategies and action plans at a range of levels.

In summary, the LEDS concept is still being developed through trials, but from what is known it is possible that there could be some concerns with the process in terms of aligning sustainable development and climate goals in the choice of sectors and technologies, ensuring a wide range of appropriate technologies and familiarity, maximising over all the benefits for a given cost and developing enabling strategies and action plans that address the whole system needs for successful acceleration innovation on a large scale.

Therefore, some merging or harmonisation of approaches, as discussed in Chap. 3, would be expected to be beneficial.

4.2.4 Comparison Across the Cancun Provision

Table 4.1 summarises the findings in this section and the previous chapters by describing commonalities and difference between TNAs, LEDS, NAMAs and NAPs. These can be considered as the provisions by which the mechanisms will achieve their goals. The summary is organised along the three key stages in a low-emission and climate-resilient policy preparation process:

1. Formulating a long-term sustainable development vision;
2. Identifying low-emission climate-resilient technologies and measures that support sustainable development; and
3. Formulating projects, programmes and strategies for policies and measures.

Figure 4.1 shows a possible representation of how the processes for achieving strategies and action plans for a low-emission future, including their interlinkages, as discussed in this section, could operate under the UNFCCC. It summarises how the processes and their outputs described in Chaps. 2 and 3 contribute to each technology innovation stage and how they relate to the new mechanisms under the Cancun Agreements (e.g., Technology Mechanism and Cancun Adaptation Framework).

The figure describes the country process for formulating accelerating strategies and action plans through, e.g., a TNA or LEDS. In Chaps. 2 and 3, we have discussed how these processes work towards objectives (e.g., portfolio of technologies or measures, inputs for NAMAs) and how this supports countries in generating innovation strategies with action plans. On the left-hand side, it is shown how the possible exchange of information and support from the Technology Mechanism and/or Cancun Adaptation Framework and Capacity-Building Framework is then related to the stages of the country process. The latter connection will be discussed in the next sections.

4.2.5 Harmonisation Across Processes

The overview of commonalities and differences between the processes for low-emission and climate-resilient development under the Convention justifies a conclusion that there is scope for harmonisation of these processes. For instance, as concluded in Chap. 3, the updated TNA process contains detailed methodologies for

Table 4.1 Overview of commonalities and differences between provisions discussed

Commonalities	Differences
a. To what extent are TNA, LEDS, NAMA, NAP embedded in a country's long term development vision?	
Most processes have a common focus *on a country's overall sustainable development context*	Unlike for TNAs, no specific methodology has been adopted yet under the UNFCCC for LEDS, NAMAs and NAPs
TNA identifies priority sustainable development strategic (sub)sectors and areas could possibly be used as inputs for LEDS, NAMAs and NAPs	
All processes are participatory	
b. How are technologies or measures for mitigation and adaptation in the country identified within this vision?	
TNA procedures are in principle suitable for the other policy concepts that identify technologies and actions in light of climate policy and sustainable development	TNAs explicitly focus on technology choices (both 'hard' or market-based and 'soft' or non-market-based technologies, such as coping strategies). In LEDS prioritisation appears to be based on cost efficiency of carbon reduction rather than maximising over all the benefits as in TNA. NAMA and NAP prioritisation of technologies has not been defined but could follow TNA
Therefore, prioritised technology portfolios from a TNA could be input for formulating LEDS, NAMAs, and NAPs	
c. What actions are envisaged to work out a pathway with lower GHG emissions or enhanced adaptation in light of the country's overall development objectives?	
There is a common focus on *strategic pathways* with action plans at the technology level (TNA), and sector and national level (TNA, NAMA, NAP and LEDS)	The updated TNA process identifies a number of core elements for a strategy, including policies and measures, whereas LEDS studies thus far seem to restrict the strategic scope to policy actions. The updated TNA process acknowledges that not all activities for accelerating innovation will be policy driven
LEDS, NAMA and NAP formulation could possibly benefit from the identification in a TNA of actions for acceleration of technologies for mitigation and adaptation	The new TNA process facilitates financial allocation for technology transfer and innovation as well as MRV but does not specifically try to locate sources of finance which LEDS appears to do

Source This table is based on a table published earlier in UNFCCC (2011)

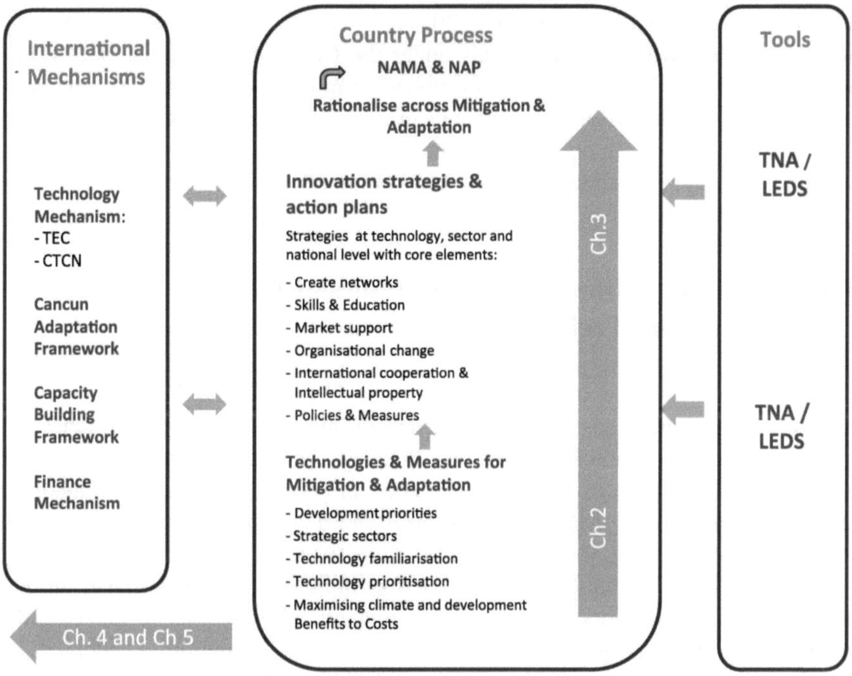

Fig. 4.1 A possible representation of how the process for achieving strategies and action plans for a low-emission future could operate under the cancun agreements

prioritising technologies and measures and formulating technology acceleration strategies. These outputs can be integrated in LEDS formulation processes, but also in the opposite direction: The policy analysis from LEDS could be incorporated in a TNA under the core element of policy and measures. As explained earlier and illustrated in Fig. 4.1, both the outputs of TNAs and LEDS would produce outputs for NAMA and NAP formulation.

Formally, there could be a reason for countries to keep TNA and LEDS processes separate. A TNA, given its origin in the UNFCCC Technology Transfer Framework, could be seen as 'politically neutral', whereas a LEDS may be considered to have a stronger connection with NAMAs (as both are included in the same chapter on 'mitigation' in the Cancun Agreements) and, therefore, be more 'political'. Technically, however, there is little reason why TNA and LEDS lessons and processes should not be harmonised or 'merged'. In spite of the different names and status under the Convention, the envisaged outputs go in the same direction.

Harmonisation of process steps makes sense not only to enable more efficient use of developing countries' resources but could also result in a clearer picture of a country's needs in terms of technologies, finance and capacity building for reaching their sustainable development goals. This would, for instance, support the work of the Technology Mechanism in terms of providing products/tools, services and partnerships to enhance the implementation of the identified actions for mitigation and adaptation (UNFCCC 2011). It would also enhance the matching of finance needs with available funds under the Convention or other bilateral or multilateral funds. These funds are discussed in more detail in the next chapter.

In the light of this discussion, we will in the remainder of this chapter refer to TNA and LEDS, as both concepts, despite their current differences, aim at similar outcomes that are embedded in countries' sustainable development context.

4.3 Supporting Climate and Development Goals Through the Technology Mechanism

Technology transfer is an important aspect of the UNFCCC and the mechanism for how it would operate has been transformed progressively as its importance for moving to a low-emission future has been realised. The Technology Mechanism is based on the activities of the former EGTT, and was established at Cancun to facilitate enhancement of technology transfer and development for mitigation and adaptation for achieving the full UNFCCC objectives.

It will have a Technology Executive Committee (TEC) that 'shall further implement the framework for meaningful and effective actions to enhance the implementation of Article 4, paragraph 5, of the Convention…. adopted by decision 4/CP.7 and enhanced by decision 3/CP.13' (UNFCCC 2010, para 119).

More specifically, the TEC will support developing countries in developing and enhancing capacity for technology RD&D, deployment and diffusion of both soft and hard technologies for mitigation and adaptation, as well as getting access to private and public investment funding. The second component of the Technology Mechanism is the Climate Technology Centre and Network (CTCN), which has a more operational focus as it 'shall facilitate a Network of national, regional, sectoral and international technology networks, organizations and initiatives' (UNFCCC 2010, para 123).

UNFCCC (2011) explores possible interlinkages between the Technology Mechanism and TNAs. These were considered from two viewpoints. On one hand, it was argued that the Technology Mechanism could support TNA & LEDS processes, such as by providing developing country stakeholders with tools, guidance and access to networks. On the other hand, the article explained how the output from TNA & LEDS could provide useful information to the Technology Mechanism in terms of insights on countries' technology, finance and capacity needs.

4.3.1 Interaction of Technology Mechanism with TNA & LEDS: Supporting the TEC to Acquire a Larger Picture

As explained in UNFCCC (2011), the TEC can use outputs from TNA & LEDS, such as NAMAs and NAPs, to obtain a larger picture of needs for technologies and other measures for low-emission and climate-resilient development in developing countries and regions (Box 4.2 presents an overview of the TEC activities). For instance, TNAs result in portfolios of prioritised technologies for mitigation and adaptation for strategic (sub)sectors, categorised in terms of availability over time and scale of application. Prioritised technologies are assessed at the level of (sub)sectors by asking the question: what climate and development benefits would the technology deliver if it were implemented in the (sub)sector at its maximum technical potential? Countries can present this information in summary tables showing technology benefits and costs (see, for example, Table 2.1 in Chap. 2). In the final stage of the TNA, the role of a technology within the (sub)sector is refined by also exploring implementation requirements for the technology in terms of overcoming the barriers in the system producing strategies and action plans for successful innovation.

Box 4.2 Overview of TEC activities

Gather and aggregate information

- Provide an overview of technological needs and analysis of policy and technical issues related to the development and transfer of technology.

Analyse information

- Promote and facilitate collaboration on the development and transfer of technology between governments, the private sector, non-profit organisations and academic and research communities.
- Seek co-operation with relevant international technology initiatives, stakeholders and organisations; promote coherence and co-operation across technology activities, including activities under and outside of the Convention.

Translate into policy

- Consider and recommend actions to promote technology development and transfer to accelerate action.
- Recommend guidance on policies and program priorities.
- Recommend actions to address the barriers to technology development and transfer to enable enhanced action.
- Catalyse the development and use of technology road maps or action plans at international, regional and national levels through co-operation between relevant stakeholders.

Source UNFCCC (2010, Chap. 4, section B)..

The TEC could synthesise this information by creating an interface where the TNA and LEDS outputs can be electronically collected so that conclusions can be drawn on, e.g., what are small-scale technology needs in Small Island Development States? what are the requirements for accelerating the implementation of

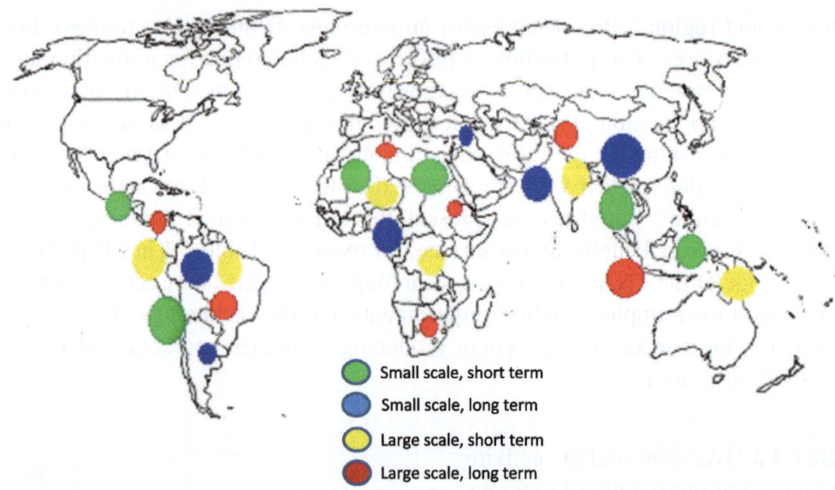

Fig. 4.2 Example of TEC Window for climate technology investments at subsector level for different regions (UNFCCC 2011, p. 10)

these technologies in the region? and what are the good practices for implementing TNA and LEDS outputs and system improvements (UNFCCC 2011)?

An example of such an interface is shown in Fig. 4.2 that presents, as part of a purely hypothetical example, regionally aggregated figures on the need for small-scale short-term technologies and measures, large-scale short-term technologies and measures, etc. The investment needs are visualised in this example by the size of the circles.

With this larger picture, the TEC would not only obtain an idea of what is needed where but also gain insight in regional capacity and finance support needs. This can be done because, in addition to summary tables with identified technology needs, the TNA and LEDS also identify actions for acceleration of technology development and transfer, which, in a synthesised form, would help the TEC to efficiently streamline and guide capacity and financial support to where it is needed.

For example, the Finance section of the Cancun Agreements contains a decision to mobilise USD 100 billion per year from private and public sources by the year 2020 (UNFCCC 2010, Chap. 4, section A). On the allocation of this money, among other provisions, the text takes 'note of relevant reports on the financing needs and options for mobilization of resources to address the needs of developing country Parties with regard to climate change adaptation and mitigation' (UNFCCC 2010, Chap. 4, section A, para 101).

The allocation of funding could thus be based on such information as individual and/or synthesised TNA, LEDS, NAP or NAMA documents.[5] Such documents show financial needs in terms of required investments and measures for creating enabling environments for technologies. With this information, the TEC could create thematic funding windows showing, e.g., required investment capital and enabling environment support for deployment and diffusion of small-scale and large technologies in different regions, specified for (sub)sectors (as shown in the example of Fig. 4.3). This information could be channelled to private, public or multilateral finance institutes who are most suitable for providing this assistance. An integrated finance strategy is discussed further in Chap. 5.

Based on the technology, finance and capacity needs assessments, the TEC can provide policy recommendations on technology development and transfer and innovation system development to the COP, as well as support the work of the CTCN to improve the products, services and partnerships for supporting developing countries with technology development and transfer. The latter includes collaboration between governments, industry, non-profit organisations, academic and research institutions bearing in mind that the mix of actors will change over time. In addition, most poverty alleviation initiatives are best started through development NGO partners who know the field before being diffused further through a mix of partners.

In light of the above, the TEC could facilitate a bigger platform to implement technology solutions that go beyond national boundaries. Information from such processes as TNA and LEDS can play a role in all these aspects. For example, the action plans for the strategies for acceleration of innovation from a TNA involve characterisation of the measures identified. An example of a training action is given in the following:

- *Why* is this measure important? For example, why is training needed?
- *How* should this training be done? Should foreign experts be invited to run a training program for domestic engineers (e.g., for a programme of small-scale short-term technology spread across the country), or should engineers from within the domestic sector be trained abroad by colleagues in developed or other developing countries (e.g., for a Integrated gasification combined cycle technology)?
- In line with the 'how?' question, *who* in the country (private and public sector entities) should be involved in arranging the training in terms of responsibility

[5] In addition to these instruments, the TEC work could also be supported by Technology Roadmaps, such as those prepared by the International Energy Agency (IEA) to advance innovative energy technology. IEA is developing a series of global low-emission energy technology roadmaps covering a range of important technologies. A roadmap starts from a particular technology and subsequently focuses mainly on milestones for successful technology RD&D, deployment and diffusion. For example, roadmaps could provide insights into legal/regulatory needs for particular technologies, investment requirements, public engagement and international collaboration. Roadmaps could be considered equivalent to the technology-level strategy from a TNA. For further information, please visit: http://www.iea.org/subjectqueries/keyresult.asp?KEYWORD_ID=4156. Accessed 28 September 2011.

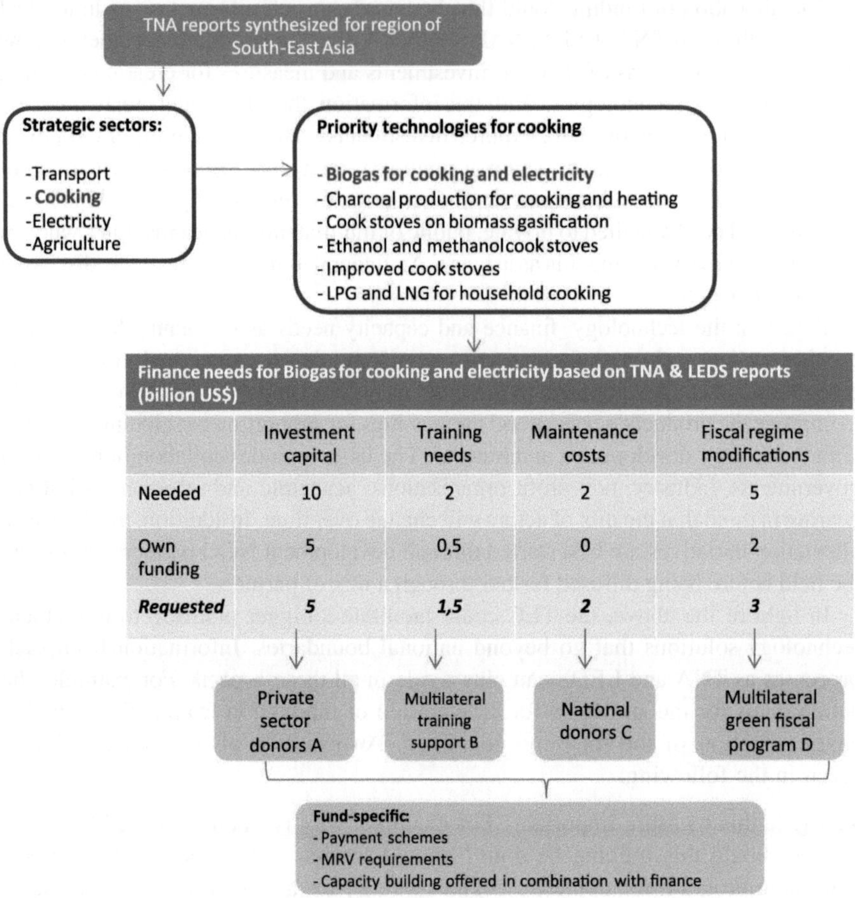

Fig. 4.3 Hypothetical example interaction TNA and LEDS—finance mechanism (UNFCCC 2011, p. 13)

and work, and with which international partners would collaboration need to be sought?

- *When* does the training need to be undertaken? In other words, what is the timeline for a full completion of the training so that domestic engineers can independently operate the technology, when is the starting date for the training and when can the training be completed? Do we also need regular courses with international experts to refresh the knowledge?
- *What* are the MRV requirements for the training to guarantee successful transfer and maintenance of knowledge? Does the country have MRV protocols available for the training programme, or can international protocols be acquired with consultancy support?
- *What costs* would be involved with the training measure?

Through the action plans (as part of the TNA and LEDS), allocation of resources and MRV of the activities can help to ensure the efficient use of financial and other resources.

In addition, the TEC might want to establish a consultative group composed of experts and representatives of organisations that are heavily involved in major international technology collaboration programmes and initiatives. Part of the TNA and LEDS process is to recognise and identify key stakeholders in the technology development and transfer process.

Finally, TNA and LEDS can provide important inputs to the agenda of the Technology Mechanism as they directly link low-emission and climate-resilient response options and acceleration of their transfers into national systems to national development priorities in developing countries. This information enables the Technology Mechanism to consider development priorities such as poverty alleviation, increased energy security of supply and improved health conditions in relation to climate policy objectives.

4.3.2 Interaction of the CTCN with TNA and LEDS

The tasks of CTCN are more operational than those of the TEC. The CTCN will have a number of functions, which are shown in Box 4.3. Broadly, these functions can be provided with the following operational modalities (UNFCCC 2010):

- Offering or guiding countries towards *tools* that they can use for making technology choices and policy preparation and implementation, such as analytical tools, informational tools and good practices;
- Offering or providing access to *services*, such as training, advice and expert teams;
- Establishing or facilitating access to international *partnerships*, such as twinning, and North–South, South–South, South–South–North co-operation; and
- Facilitate supporting accelerating activities for innovation, e.g., new networks and access to existing networks, policy implementation support, market support and enforcement of activities.

Box 4.3 Functions of the Climate Technology Centre and Networks (CTCN)

1. Provide advice and support at the request of developing countries related to the identification of technology needs and implementation of environmentally sound technologies, practices and processes.
2. Support developing countries' capacity to identify technology options, make technology choices and operate, maintain and adapt technologies to local circumstances.

3. Facilitate prompt action on the deployment of existing technologies in developing countries.
4. Stimulate and encourage development and transfer of existing and emerging low-emission and climate-resilient technologies, possibly by different forms of international country co-operation, such as North–South and South–South–North co-operation.
5. Facilitate international partnerships among public and private stakeholders for acceleration of innovation and diffusion of technologies.
6. Provide in-country training to support technology actions in developing countries.
7. Identify, disseminate and assist with developing analytical tools, policies and best practice for country-driven planning for technology development and transfer.

Source (UNFCCC 2010, para 123 a–d.)

With these functions and operational modalities, the TNA and LEDS processes can offer direct support to a CTCN in terms of supporting identification of 'hard' and 'soft' technology needs by providing tools for, e.g., participatory processes; data collection and quality check in the different steps of a needs assessment; technology familiarisation, e.g., *ClimateTechwiki* (see Chap. 2); multi criteria decision analysis, such as *TNAssess* (see Chap. 2); market mapping (see Chap. 3); and sharing of good practices. Also the access to international partnerships through the CTCN might help an individual country during the different steps of an assessment, such as support for understanding the potential impact of climate change on a country's economic sectors and need for adaptation, as well as sharing the experience on creating an enabling environment for prioritised technologies or practices with formulation of projects, programmes or strategies (as referred to in UNFCCC 2010, para 123 (a) i–iii). This could also support developing countries to formulate NAMAs and NAPs based on technologies and strategies formulated in TNA and LEDS and take prompt action on these outputs (UNFCCC 2010, para 123 (a) iii).

However, as explained in UNFCCC (2011. p. 11), outputs of TNAs and LEDS could also 'inform the design of the CTCN and its evolution over the time in terms of countries' changing technology needs'. For instance, the CTCN could undertake a more pro-active role 'in terms of, e.g., knowledge exchange and matching developing countries' technology, financial and capacity-building needs with multilateral or bilateral capacity support and funding resources. Based on this matching, a clearer picture can be obtained of opportunities and gaps in the fields of technologies, finance, and capacity building to expand and enhance existing support' (UNFCCC 2011, p. 11).

This role could be supported by using insights from concluded and ongoing TNA and LEDS for further development and customisation of tools ('What are

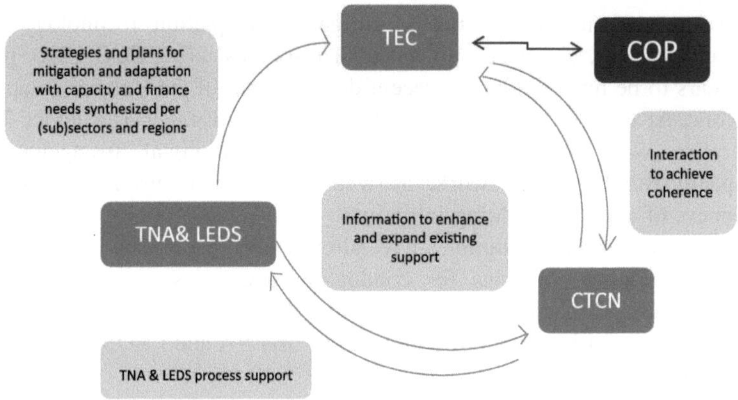

Fig. 4.4 Overview of relationship TNA and technology mechanism (UNFCCC 2011, p. 12)

generally requested tools?') and advice and for enhancing partnerships under the CTCN. Insights from needs assessments 'may also provide guidance on what could be reasonably and effectively expected from the CTCN in terms of supporting the acceleration of technology development and transfer in developing countries' (UNFCCC 2011, pp. 10–11).

The possible interlinkages between TNA and LEDS and the role of the Technology Mechanism and its two bodies, as discussed in this section, are summarised in Fig. 4.4.

In the preceding section, it was discussed how the TEC and CTCN could be designed to support developing countries in conducting low-emission and climate-resilient needs assessments and in implementing the results. It was also illustrated how outputs from needs assessments, synthesised across countries, could support the TEC in presenting a larger picture of regional needs for technologies and measures for mitigation and adaptation with related capacity building, finance needs and MRVs for these. This information could form relevant input for the work under the Finance Mechanism of the Convention, which was illustrated with the example of Fig. 4.3, and discussed further in Chap. 5. Inputs for specifying capacity building support to developing countries are also produced.

4.4 Possible Interlinkages Between TNA and LEDS and Capacity Building for Mitigation and Adaptation in Developing Countries

To strengthen capacity building under the Convention, a *Capacity-Building Framework* was first established at COP 7 in Marrakech (UNFCCC 2001). The Cancun Agreements acknowledge capacity building as 'cross-cutting in nature and an integral part of enhanced action on mitigation, adaptation, technology development and transfer, and access to financial resources' (UNFCCC 2010, p. 20). This appears to be mirrored by the recent development of the Cancun Adaptation Framework. At the COP in Durban (South Africa, November–December 2011), consideration was given to the structure and institutional arrangements of the Capacity-Building Framework, as well as ways to introduce monitoring effectiveness of capacity-building initiatives.

In the previous section, examples have already been given of possible capacity needs in developing countries for conducting needs assessments (e.g., data, training and tools). It has also been argued that the CTCN can facilitate access to capacity elements for implementing the outputs of TNA and LEDS (e.g., projects, programmes, strategies, etc.).

The TNA and LEDS could also identify capacity needs through the activities identified in the strategies and action plans. To provide a picture to potential international partners of what these capacity needs look like, measures for accelerating technology innovation (as discussed in Chap. 3) can be structured into core elements for a strategy.[6] Many of these measures could be considered as capacity-building actions in the country. Characterising the measures as a part of a strategy action plan shows why the measure is important, who should do it, how, when, what are MRV requirements and how much it would cost (see the example in Sect. 4.3.1).

Such characterised measures can indicate what the country can do itself and for which parts it needs international support (see Fig. 4.3). It also provides a clear structure for identifying such capacity-building indications from the country, which can be specified, depending on what the country wants, for individual technology projects, sector programmes and national innovation strategies.

Similar to the overview windows for technology investment costs and benefits for (groups of) developing countries shown in Fig. 4.2, overviews can be prepared for capacity-building needs. For example, in region A, there is a need X for training programmes for sustainable cook stoves in rural areas. Based on this information, the TEC can translate these overviews into policy recommendations on international capacity-building support to developing countries, possibly fine tuned for particular groups of developing countries or sectors. This could also be done for adaptation measures.

From this it can be concluded that conducting TNA and LEDS could contribute to many of the capacity-building points of action mentioned in the negotiation text (see Box 4.4) in the following ways:

- It involves stakeholders in all stages of the assessment and supports overall familiarity of the country's decision makers with technologies for mitigation and adaptation. The process also enhances decision makers' feeling for linking the technology choice process with overall development priorities and facilitates an

[6] Such as policies and measures, education and training, behavioural change, market or system supporting measures, network creation and support and international co-operation and intellectual property rights, as discussed in Chap. 3.

early 'buy in' of stakeholders in the eventual implementation of the prioritised technologies.

- It supports climate policy action development in the country by offering a strategy based on prioritised technologies with an action plan for acceleration of technology innovation for mitigation and adaptation.
- It enhances domestic capacity for resource allocation and MRV work as needs assessment processes result in identified MRV requirements for enhancing the performance of prioritised technologies and successful implementation of measures for their acceleration as part of the national strategy and action plan.

Box 4.4 Capacity-building points of actions from Cancun Agreements

- Strengthening endogenous capacities, skills and capabilities;
- Development and strengthening of national and/or regional networks;
- Strengthening of climate change communication and public awareness;
- Encouraging participatory and integrated approaches in relevant social, economic and environmental policies and actions;
- Enhancing capacity to monitor and report on climate change action; and
- Enhancing capacity to plan, prepare and implement climate change actions.

Source UNFCCC (2010, Chap. 4).

As the capacity-building actions form part of the innovation strategy and action plan and are essential for accelerated transfer of prioritised technologies for mitigation and adaptation, *it would not be recommended that these should be dealt with by a body separate from the Technology Mechanism.* Capacity-building support provided for technology development and transfer by separate bodies (e.g., Technology Mechanism and Capacity-Building Framework) runs the risk that support actions are formulated and provided independently, with risk of duplication of work and fragmentation of coherent action. Integrating such activities, as is discussed in the following, is essential to mitigate this risk.

4.5 Delivering the Promise of the Cancun Agreements

4.5.1 Potential for Coherent Action Under Cancun Agreements

As a result of the Bali Plan of Action (UNFCCC 2008), negotiations have taken place with a focus on a number of building blocks or pillars for a new climate policy regime. The pillars are mitigation, adaptation, technology transfer and finance. Although in the Cancun Agreements capacity building is described as cross-cutting, from the discussion in this and earlier chapters it has become clear that also mitigation and adaptation activities could overlap and that there are cross-cutting issues for these building blocks as well.

For example, there are overlaps in areas such as water resource use and in facilitating energy access for the poor to alleviate poverty as discussed in earlier chapters. Furthermore, countries may not have either just mitigation needs or just adaptation needs, so that initiatives developed under, e.g., the Technology Mechanism and Cancun Adaptation Framework, could interfere with each other or duplicate activities, as they may not be aware of the wider range of activities in the country. This could lead to wastage in terms of funding allocation. An additional concern is that the processes undertaken by different institutions will diverge over time to the extent that syntheses will become difficult and meta-analysis at the regional or global scale of doubtful value.

As a consequence, there is a need to ensure under the Cancun Agreements an *integrated approach* on mitigation, adaptation, capacity building, technology transfer and finance so that coherent actions can be formulated and separation of outputs to different institutions avoided. Coherence in support measures and allocations would be crucial in the light of the need for urgent large-scale deployment of technologies and measures for mitigation and adaptation. The potential for integration of actions is discussed further in the following.

Without some form of integration across these building blocks or pillars, strategies and action plans for developing countries could be envisaged under the Technology Mechanism for mitigation and adaptation, but also under the Cancun Adaptation Framework for adaptation. Therefore, depending on the final structures of these mechanisms, without integrated action, there is ample scope for duplication, especially if capacity building is taken into account. As explained earlier, capacity-building actions are inherent in the technology transfer activities and innovation acceleration strategies.

In conclusion, for maximising all possible benefits from investments, it is recommended that the strategies and action plans for each country have to be treated as an integrated whole with support for each of the steps in the process, so that the country moves from development of actions to implementation as all the actions are required for maximising the potential for successful transfers. If the actions are split across separate bodies, there are risks of both failing to generate and implement the action plans coherently (e.g., with capacity-building innovation system support activities in the relevant timeframe for the technology deployment and diffusion) as well as the risk of generating conflicting or overlapping strategies, action plans and allocations.

4.5.2 Integration Options

Based on the preceding discussion, we can identify the following options for how the activities of the Technology Mechanism, Cancun Adaptation Framework and Capacity-Building Framework can be integrated, moving from no integration to full integration:

1. It is possible that the Technology Mechanism could act alone for development of strategies for acceleration of innovation for both mitigation and adaptation. The area of operation of the Cancun Adaptation Framework and Capacity-Building Framework would not be related to the large-scale urgent transfers of technologies and measures under the Technology Mechanism.
2. Alternatively, appropriate parts of the Cancun Adaptation Framework and Adaptation Committee could be merged with the Technology Mechanism to strengthen expertise in this area and similarly for the Capacity-Building Framework.
3. Another alternative is to subsume all three mechanisms into a new overarching body and put appropriate procedures and institutional structures in place to ensure that all pillars function as a whole, sharing expertise and providing support to countries to contribute to final outcomes.

Whichever option is chosen, it should be as simple and efficient as possible to avoid becoming a blockage itself in the transfer system. Teams could be formed for countries/regions to work together across mitigation, adaptation and capacity-building support.

However, there will always be a need to compare across these areas to avoid duplication and maximise benefits. Integration, therefore, also requires not just institutional but procedural integration and an additional function will be needed for mitigation, adaptation and capacity-building strands.

This leads us to the following key messages:

- It is suggested that when mitigation and adaptation strategies have been developed at project, sector or national level, then a rationalisation step at the country level or international level or both is needed to ensure coherence across the mitigation and adaptation activities and avoid wastage of resources.
- As capacity-building actions are so integrated into the transfer process, they should not be separated out or implemented separately, but kept together and managed within the strategy and action plan as a whole under the Technology Mechanism. Again, checks will be needed to ensure duplication of capacity-building activities is avoided.
- This again argues for an integrated structure for supporting low-emission technology transfer and accelerated innovation.

The preceding conclusions on the need for integration and rationalisation of actions at both the national and international level are summarised in Fig. 4.5.

In this chapter, the need for integration of technology transfer, mitigation, adaptation and capacity-building work under the Bali Action Plan and Cancun Agreement has been discussed. In the next chapter, issues related to the finance of low-emission and climate-resilient technologies and measures are discussed in the light of these insights.

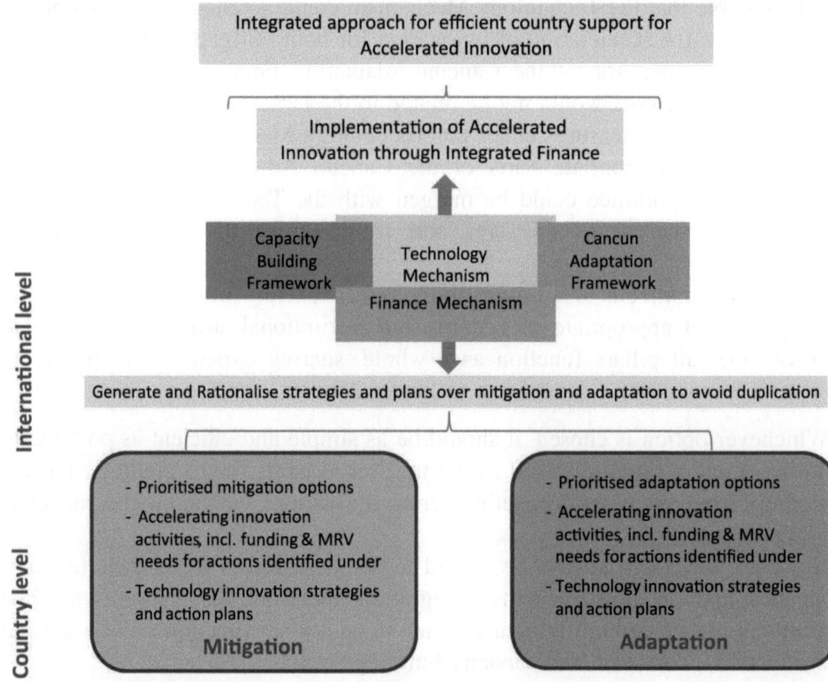

Fig. 4.5 Rationalisation of actions for mitigation and adaptation

4.6 Meeting the Challenge: Supporting Strategies for Climate and Development in a Climate Policy Regime

The chapter has discussed the challenge of how the insights from earlier in this book could be incorporated in a future climate policy regime to achieve the urgent innovation of low-emission technologies into country systems on the scale and timescale required. Recent developments in the negotiations, since the 2007 Bali Plan of Action (UNFCCC 2008), have shown an increasing focus on developing strategies for meeting climate goals in the context of countries' sustainable development. Provisions included for that in the Copenhagen Accords and Cancun Agreements are TNAs, LEDS, NAMAs and NAPs. In addition, we have demonstrated how the recently updated TNA process under the Convention nowadays supports developing countries in formulating a strategy for technology transfer and innovation, including action plans for strategy implementation. We have also discussed possible interlinkages between these provisions, such as how output from, for instance, TNA and LEDS could be used as input for NAMA or NAP formulation.

It can, therefore, be concluded that with TNA, LEDS, NAMAs and NAPs, a future climate policy regime will have important provisions and mechanisms for supporting countries in formulating long-term pathways for realising environmental, social and economic development goals with low-emission and climate-resilient technologies or other measures. We have also shown that there is a need for some harmonisation between TNA and LEDS.

In addition to these provisions, the climate negotiations have also resulted in the Technology Mechanism, the Cancun Adaptation Framework, the Capacity-Building Framework and the Finance Mechanism to support international technology development and transfer, provide adaptation, capacity building and finance support, respectively. We have argued in this chapter:

- How conducting TNA and LEDS, as a complete exercise involving technologies and their innovation acceleration activities, could support the Technology Mechanism and/or Cancun Adaptation Framework and Capacity-Building Framework, both in terms of generating strategies and action plans, but also in terms of preparing a synthesis of results for regional insights.
- How, in turn, the Technology Mechanism and Cancun Adaptation Framework could support the implementation of TNA and LEDS strategies and action plans.

Recommendations for the efficient operation of these three mechanisms in the Cancun Agreements for supporting the development and implementation of low-emission strategies and action plans have been provided. We have pointed out that there could potentially be duplication and wastage of resources, especially as the Technology Mechanism has a remit to cover both mitigation and adaptation, whereas the NAPs are mainly the responsibility of the Cancun Adaptation Framework. Thus, options for their integration have been suggested to ensure that mitigation and adaptation expertise is shared.

If these mechanisms operate independently from each other, then the overall result could be fragmentation of technology transfer strategic and support actions that in practice need to be integrated and co-ordinated. Separate actions could lead to conflicting or overlapping outputs and to isolation of mitigation, adaptation and capacity-building expertise in silos.

In addition, a new step is recommended in the country process for mitigation and adaptation strategies. This step would contain a rationalisation of mitigation and adaptation strategies and action plans to avoid overlaps and unintended duplication or interference between them. Similarly, capacity-building activities identified for acceleration of innovations cannot be separated. It is, therefore, recommended that these capacity-building activities for mitigation and adaptation remain within the Technology Mechanism to avoid unnecessary complications. Again, overlaps with the activities under the Capacity-Building Framework should be checked and rationalised.

Decisions will, therefore, be required for:

1. Harmonisation of the new TNA process and LEDS.
2. Whether the TNA and LEDS process provides NAMAs and NAPs and, therefore, avoids the need for developing countries to do four separate exercises.

3. Whether mitigation, adaptation and capacity building can be integrated in some form, specifically to support countries in producing and implementing TNA and LEDS strategies and action plans on the scale and urgency required.
4. Whether to introduce an additional rationalisation process across mitigation, adaptation and capacity building.
5. Whether to introduce an additional rationalisation step across the country mitigation and adaptation strategies and plans before submission for funding.

These conclusions are intended to support negotiators to address the challenge of organising the provisions of TNA, LEDS, NAMAs and NAPs under the Technology Mechanism, Cancun Adaptation Framework and Capacity-Building Framework in such a manner that:

- Institutions and processes are efficiently conducted and where possible in a harmonised way; so that;
- The outputs for mitigation and adaptation strategies and action plans will deliver the sustainable development benefits for the country;
- Supported with advice, access to networks, finance and overall capacity building, such as education, supporting economic, legal and technical services; and
- Improved enabling environments for adoption of measures for mitigation and adaptation in the timescales and scale required.

References

CCAP (2009). *Nationally appropriate mitigation actions by developing countries: Architecture and key issues*. Washington, DC, USA. Center for Clean Air Policy. Retrieved September 28, 2011 from http://www.ccap.org/docs/resources/823/NAMAs%20by%20Developing%20Countries%20-%20Architecture%20and%20Key%20Issues.pdf

ClimateFocus (2011). *Design options for NAMAs and their regulatory framework*. Briefing Note. Retrieved September 27, 2011 from http://www.climatefocus.com/documents/files/design_options_for_namas_and_their_regulatory_framework.pdf

Jung, M., Eisbrenner, K., Escalante, D., de Vos, R., Höhne, N. (2011). *Cancun results pave the way for nationally appropriate mitigation actions [NAMAs]*. Ecofys Policy Update Issue II. April 2011. Retrieved September 28, 2011 from http://www.ecofys.com/com/publications/brochures_newsletters-/documents/Ecofys_Policy_Update_II_04_2011.pdf

UNFCCC (2001). *The Marrakech Accords, capacity building in developing countries (non-Annex I Parties)*. Decision 2/CP.7. FCCC/CP/2001/13/Add.1. Retrieved September 29, 2011 from http://unfccc.int/resource/-docs/cop7/13a01.pdf#page=5

UNFCCC (2004). *Buenos Aires programme of Work on Adaptation and Response Measures*. Decision 1/CP.10. Retrieved September 28, 2011 from http://unfccc.int/files/meetings/cop_10/adopted_decisions/application/pdf-/01_cp_l_16.pdf

UNFCCC (2007). *Report of the Conference of the Parties on its twelfth session, held at Nairobi from 6 to 17 November 2006*. FCCC/CP/2006/5/Add.1. Retrieved September 28, 2011 from http://unfccc.int/resource/-docs/2006/cop12/eng/05a01.pdf#page=13

UNFCCC (2008). *Report of the Conference of the Parties on its thirteenth session, held in Bali from 3 to 15 December 2007*. FCCC/CP/2007/6/Add.1. Retrieved September 28, 2011 from http://unfccc.int/resource/-docs/2007/cop13/eng/06a01.pdf

UNFCCC (2010). *The Cancun Agreements: Outcome of the work of the Ad Hoc working group on long-term cooperative action under the convention.* Decision 1/CP.16. FCCC/CP/2010/7/Add.1. Retrieved July 17, 2011 from http://unfccc.int/resource/docs/2010/cop16/eng/07a01.pdf#page=2

UNFCCC (2011). *Interlinkages between technology needs assessments and national and international climate policy making processes*. Background paper III for UNFCCC workshop on technology needs assessments, Bonn, Germany, 1–2 June 2011. Retrieved June 3, 2011 from http://unfccc.int/ttclear/pdf/Workshops/Bonn%202011/Background%20Paper%20III%2025%20May%20Final.pdf

Chapter 5
Challenge 5: Financing Technologies and Actions for Climate and Development

Abstract This chapter discusses the challenge of how to finance the various activities for a pathway for meeting climate and development goals in the time-scale and at the necessary scale. This challenge is addressed by providing an overview of the overall finance needs estimates for climate change mitigation and adaptation as well as existing finance initiatives and pledged funding under the Cancun Agreements. The chapter then discusses experience with financing activities for realising development benefits, such as energy access for poor people. This is followed by an analysis of how finance initiatives for the purchase and modification of technologies being transferred for climate and development benefits can be scaled up, as well as how to scale up finance initiatives for technology acceleration actions as part of a strategy, such as policies, enforcement instruments, network building, training, supply chain development, etc. Finally, the chapter explores how, with the large number of private and public players identified, a finance strategy can be generated and integrated with the TNA & LEDS innovation strategies to maximise successful transfers.

5.1 Challenge 5: Introduction

A final challenge to be discussed in this book is how to finance the various activities that countries would identify for meeting climate and development goals. In Chap. 2, it has been explained how a technology prioritisation process, such as the Technology Needs Assessment (TNA), would result in portfolios of technologies for mitigation and adaptation within (sub)sectors and within different categories (short and long term, small and large scale). In Chap. 3, discussion focussed on how actions can be identified by stakeholder groups to accelerate the development, deployment and diffusion of these technologies. These actions form the strategies at the technology level. Aggregating and rationalising these actions

W. van der Gaast and K. Begg, *Challenges and Solutions for Climate Change*, Green Energy and Technology, DOI: 10.1007/978-1-84996-399-2_5,

across technologies could lead to (sub)sector-level strategies, and aggregation across (sub)sectors could result in national strategies.

In both Chaps. 2 and 3, it has been demonstrated how stakeholders in a technology prioritisation and strategy formulation process can identify finance needs for technologies and for strategic actions.

For example, in a TNA, the development and climate benefits and costs of technologies are compared to explore which technologies offer the largest overall benefits to the costs. In the summary tables for prioritised technologies in a TNA, benefits and costs are estimated for transfer into a sector over a longer time period (e.g., for 20 years), including investment and operational and management costs of technologies. Subsequently, actions for acceleration of technology development and transfer within a country are characterised for technology, sector and national strategies (e.g., why is it important, who is responsible, when should it be done, etc.).

Part of this characterisation is how much money will be needed to undertake the actions (e.g., a training programme for an improved cooking stove programme in the country could cost US$ 2 million for the next 5 years). Therefore, the process provides costs for the technologies or measures and for the innovation actions to accelerate their adoption. Similar data would be expected in a LEDS.

Experience of deployment of technologies is currently largely focussed at individual projects or small-scale programmes. Therefore, a change in scale will be needed so that the technologies or measures can be rolled out at the desired level for reaching climate and development goals.

To achieve successful acceleration of innovation of low-emission and climate-resilient technologies at the required scale, it was pointed out in Chap. 3 that new technologies and the strategies to accelerate their adoption in the country should be implemented together.

This means that consideration has to be given to scaling up both funding for technologies and to funding for innovation support actions that will be needed on the same scale. This represents the main challenge to be addressed in this chapter:

> To facilitate development of an international framework for financial resources to support technology development and transfer, improve enabling environments for innovation, address equity issues such as poor people's energy access and make implementation of activities possible at the desired scale within the country.

To address this challenge, it will be divided into different sub-challenges that are discussed further in the next sections, as follows:

1. Scaling up finance initiatives for achieving mitigation and adaptation goals.
2. Creation and scaling up finance initiatives for realising development benefits, such as energy access for poor people and their access to other basic needs. Lessons from development aid experience are discussed.
3. Scaling up finance initiatives for the purchase, modification, construction and installation of technologies being transferred for climate and development benefits.
4. Creation and scaling up of finance initiatives for actions identified for acceleration of technology development and transfer as part of a strategy, such as policies, enforcement instruments, network building, training, supply chain development, etc., to ensure successful transfers into the country system.

These challenges are discussed in the next sections, starting with an overview of funding needs for mitigation and adaptation. This is followed by a discussion on experiences with supporting universal energy access and aid effectiveness and how these relate to low-emission and climate-resilient development. Finally, we discuss together the funding needs for technology equipment and technology acceleration actions and how private and public finance initiatives could fulfil these.

5.2 Scaling Up Finance Initiatives for Mitigation and Adaptation

Several recent studies have estimated how much funding would be needed to achieve the climate policy target of limiting global average temperature increases to 2°C, as well as to support developing countries in adapting themselves to climatic changes. These studies often aim at presenting a larger picture of financial needs, sometimes specified for regions and sectors. However, as will be discussed elsewhere in this section, the studies address the costs of new technologies and measures, but generally do not address funding needs for accelerating and supporting the innovation of the technologies and measures into the country systems, though some infrastructure costs are included.

The section also provides an overview of existing finance initiatives for mitigation and adaptation for developing countries, thereby distinguishing between governmental funding through bilateral co-operation between countries, funding provided by multilateral organisations (e.g., regional development banks) and funding generated by international carbon markets.

5.2.1 Overall Finance Needs Identified

5.2.1.1 Finance Needs for Mitigation

Estimated investment needs for realising a pathway to help the world to stay below the 2°C target (IPCC 2007, UNFCCC 2010) have been presented in a range from 55 to 80 billion each year[1] during the period 2010–2020 (Project Catalyst 2010), to US$ 400 billion between 2010 and 2030, rising to more than US$ 1 trillion from 2030 to 2050 (IEA 2010). According to UNFCCC (2007), global additional investment and financial flows in 2030 need to amount to US$ 200 to 210 billion to return global GHG emissions to 2007 levels.[2]

Project Catalyst (2010) made an interesting comparison between investment needs for low-emission development and total infrastructure investment. They conclude that while the latter is estimated at US$ 7 trillion per annum by the year 2020, US$ 1.5 trillion of that amount is energy-related and US$ 290 billion per year will be needed by 2020 for capital investment for low-emission infrastructures in developing countries. According to Project Catalyst (2010), US$ 215 billion of that amount is likely to be spent in 'wealthier developing countries'. In terms of sector division, Project Catalyst (2010) expects most of these investments in developing countries' power sector (US$ 155 billion per year) and energy efficiency measures (US$ 123 billion per year).

In the National Economic, Environment and Development Study (NEEDS), the UNFCCC assessed mitigation and adaptation financing needs for 11 developing countries[3] (UNFCCC 2011a). It was found that country-level mitigation costs could range from US$ 31 to 45 million, depending on the country in question and the time-horizon taken (2030 or 2050). Table 5.1 presents the findings for a number of study countries from different regions and with different sizes and economic contexts.

5.2.1.2 Finance Needs for Adaptation

UNFCCC (2007) concludes that global additional investment and financial flows needed are in the order of 'several tens of billions United States dollars', noting that the global cost of adaptation is difficult to determine. The UNDP 2007/2008 UN Development Report (UNDP 2007) estimates that 'at least US$ 40 billion will be needed by 2015 to strengthen national strategies for poverty reduction in the

---

[1] In addition to Official Development Assistance (ODA) and Clean Development Mechanism (CDM) funding.

[2] In terms of global GDP levels, this funding would amount to 0.3–0.5% or 1.1–1.7% when compared with global investment levels (UNFCCC2008).

[3] These countries are Costa Rica, Egypt, Ghana, Indonesia, Jordan, Lebanon, Maldives, Mali, Nigeria, Pakistan and the Philippines.

Table 5.1 Examples of aggregated estimated short-term (2030) and long-term (2050) financial costs of mitigation (costs in billion US$)

Country	Short-term costs	Long-term costs	Sectors covered
Egypt	4.2	12.6	Energy sector
Ghana	0.3406	0.4227	Energy, forestry, transportation
Jordan	8.5[a]	7.84[b]	Energy and waste
Mali	11.01	33.01	Forestry, agriculture
Nigeria	0.874 (annually)	1.41 (annually)	Energy (to reduce GHG emissions by 25%)
	2.9	4.8	Afforestation
	1.78	2.98	Agro-forestry
	0.444	0.889	Establishing forest units

Source adapted from UNFCCC 2011a

[a] This figure represents the total cost of identified mitigation projects in the energy and waste sectors up to 2020. Of this amount, incremental costs of US$ 3.22 billion for the energy sector and USD 125 million for the waste sector were identified.

[b] This figure represents only the incremental costs of potential mitigation measures up to 2050.

face of climate change risks', whereas by the same year, 'at least US$44 billion will be required annually for "climate proofing" development investments (2005 prices)'. However, as pointed out by ActionAid (2009), 'if significant emission reductions are not achieved in the short term, these figures will only increase as rising global temperatures generate worsening impacts'.

Comparison of estimates of overall finance needs for adaptation is difficult as several studies have used different definitions of what is meant by adaptation. For example, some studies include cost of adapting poverty reduction strategies to climate changes, climate proofing and strengthening disaster responses, whereas others identify a broad range of sectors for adaptation and estimate costs of equipment, measures and infrastructure improvements. As a result, finance needs estimates for adaptation could range from US$ 37 in Stern (2007) to US$ 171 in UNFCCC (2007).

Finally, as is discussed later in this chapter, adaptation is strongly related to people's ability in developing countries to achieve development goals. For example, as discussed by Anderson (2011), to make the Millennium Development Goals in Africa climate resilient, 40% more funding will be required during the next 10 years in comparison to a scenario without climate change. In other words, for achieving development goals, additional funding will be needed as a result of adaptation to climate change needs. This example clearly illustrates how inter-linkages between climate change and sustainable development goals, as discussed in Chap. 1, also affect the financing needs for developing countries.

5.2.1.3 Discussion

Although the above estimates are important to acquire a general sense of the scale of investments needed, it is also important to interpret them with care. First, the estimates are often expressed on a highly aggregated level and do not reveal country or region-specific finance needs, such as in the case of adaptation. Second, estimates could be difficult to compare because of differences in methodologies used, uncertainty about data used (especially in the case of developing and least developed countries) and because they are often based on scenario analysis with incorporation of (marginal) costs of technology options, but without cost figures for actions for acceleration of technologies within a country's markets or systems (as discussed in Chap. 3).

As discussed in Chap. 4, insights on overall finance needs for technologies and measures for mitigation and adaptation, as well as for actions for implementing technology innovation strategies, could also be derived from assessments such as TNA and LEDS. This could be done by aggregating such finance needs across countries and possibly across regions.

In Chap. 4, it was recommended that this activity, as a part of an overall funding strategy for mitigation and adaptation, be carried out under the Technology Mechanism in an integrated approach with the Cancun Adaptation Framework, Capacity-Building Framework and Finance Mechanism under the Convention. Part of this funding strategy would be to identify available finance initiatives for climate and development and their suitability for funding equipment and actions for innovation strategies. This will be further discussed later on in this chapter.

Before that, we first present an overview of presently available funding under the UNFCCC.

5.2.2 Available Funding Under the UNFCCC for Mitigation and Adaptation

5.2.2.1 Financial Pledges Under the Cancun Agreements

According to the UN High-level Advisory Group on Climate Financing (AGF), concerted efforts and mobilisation of both public and private sector financial resources to make available sufficient financial sources is 'challenging but feasible' (AGF 2010).[4]

An important step in the process of financially supporting developing countries to progress with achieving climate (especially adaptation) and development goals was made at the Copenhagen and Cancun Climate Conferences (COP) of December 2009 and December 2010. As a part of the Cancun Agreements,

[4] Public finance could be defined as (a) domestic-fiscal measures, public subsidy schemes, export subsidies, import tariffs, production quota systems, etc.; (b) international, regional funds/programmes (bi- or multilateral), official development aid, etc. Private financial sources are, for instance, corporate (inter)national finance, private equity, (semi-)commercial loans (including micro-credit), market-based instruments (e.g., CDM, tradable certificates, eco-labels, etc.), philanthropy/charity, etc.

developed countries made three key pledges on financial transfers to developing countries for mitigation and adaptation measures (UNFCCC 2010, Chap. 4, A. Finance):

1. To provide resources approaching US$ 30 billion in fast-start finance for the period 2010–2012;
2. To mobilise US$ 100 billion per year from private and public sources by the year 2020; and
3. To establish a new Green Climate Fund as an operating entity of the financial mechanism under the Convention (see Box 5.1).

These pledges were made within the context of the UNFCCC Financial Mechanism (UNFCCC 1992 Art. 11.1), which was discussed in Chap. 4.

Box 5.1 Green Climate Fund

One key outcome of the Cancun Agreements was the establishment of the Green Climate Fund. This fund, designated as an operating entity of the UNFCCC's Financial Mechanism (in addition to the Global Environment Facility, GEF), is to 'support projects, programmes, policies and other activities in developing country Parties using thematic funding windows' (UNFCCC 2010, Chap. 4, A. Finance). According to the Cancun Agreements, funding for the fund 'may come from a wide variety of sources, public and private, bilateral and multilateral, including alternative sources'. However, the specifics of the Green Climate fund are still to be fleshed out during the negotiations (officially through the Transitional Committee).

The fast-start finances of USD 30 billion are intended to support immediate action on climate change and signal the commitment of developed countries to developing countries in assisting them in their transition. Table 5.2 presents an overview of fast-start finance pledged by industrialised countries and what has already been committed by mid-2011, as reported to the UNFCCC secretariat. As can be seen from the table, by mid-2011, the amounts pledged by countries had almost reached the target of US$ 30 billion, whereas US$ 12.56 billion had been committed.

The Cancun Agreements also provide for the establishment of a Standing Committee (see also Chap. 4), which is to 'assist the Conference of the Parties in exercising its functions with respect to the financial mechanism of the Convention' (UNFCCC 2010, para 112). Its main areas are:

(a) Improving coherence and co-ordination in the delivery of climate change financing;
(b) Rationalisation of the Financial Mechanism;
(c) Mobilisation of financial resources; and
(d) Measuring, reporting and verifying (MRV) support provided to developing country parties.

Table 5.2 Fast-start finance as pledged and/or committed by several countries (in US$ million)

Country	Pledged	Committed
France	1,804	601
Belgium	215	57
Denmark	231	53
Finland	157	35
European Commission	215	72
Germany	1,804	510
Iceland	1	1
Ireland	143	–
Luxembourg	13	4
Malta	1	0.2
Netherlands	444	–
Portugal	52	17
Slovenia	11	0.7
Spain	537	192
Sweden	1,145	165
U.K.	2,454	929
Other EU countries	1,295	–
Australia	640	640
Canada	414	–
Japan	15,000	7,200
Norway	1,000	382
Switzerland	159	
USA	1,704	1,704
Total	28,144	12,562

Source WRI (2011)

As explained in Chap. 4, through the activities of the Standing Committee, the financial support to developing countries can be better streamlined. Such a task would be in line with the suggestion in Chaps. 3 and 4 that at a certain point in the process of formulating technology innovation strategies for a sector or country (e.g., through TNAs or LEDS), a rationalisation step is needed, including for identified financing needs (see Fig. 4.3 in Chap. 4 for a hypothetical example of such a step). With that, overlaps between mitigation and adaptation technology strategies, as well as gaps, can be identified and dealt with.

Finance will also be essential to support MRV development for a country to justify how the funding was spent and what have been the results (see Chap. 4). This support could be especially important for developing countries when attracting private sector funding; a country with proper MRV structures in place is likely to attract more private finance for its climate and development investments.

The pledges made at Cancun cover 'new and additional' finances that will be directed to developing countries (UNFCCC 2010, para 97). The Cancun Agreements highlight that funds may come from a wide variety of sources (see Box 5.1). With this, the Cancun Agreements seem to acknowledge the need for rapidly

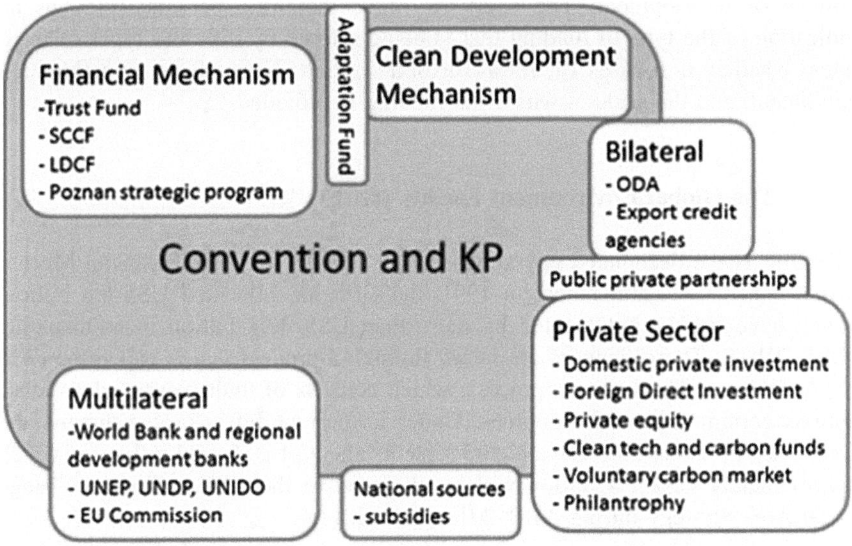

Fig. 5.1 Summary of potential financial resources under the convention and Kyoto Protocol for developing countries (*Source* UNFCCC 2011b)

expanding private investments to make progress on achieving climate and development goals, including acceleration of technology development, deployment and diffusion (Nassiry and Wheeler 2011).

Figure 5.1 presents an overview of different types of funds from which financial resources can be channelled to developing countries. They fall into four categories:

(a) Public finance through the United Nations (especially the Financial Mechanism and the operating entity GEF);
(b) Bilateral, multilateral and regional funding sources such as donor countries and the regional development banks (e.g., Asian Development Bank or the African Development Bank);
(c) Carbon markets (e.g., the Clean Development Mechanism (CDM)); and
(d) Private sector investments.

A substantial amount of funding has been made available already by these sources, both individually and in combination with other financial sources (e.g., public–private sector partnership). In the next sections, further detail is provided on how the funding has been generated by describing the supporting activities of the GEF, bilateral, multilateral and regional funding sources, and the carbon market funding.

It must be noted though that the overview in the next sections is not meant to be exhaustive as there are many funds active nowadays in the field of development and climate and for several funds, it is even difficult to classify them as focusing on climate or development. The overview must, therefore, be considered as an indication of the type of funding that is being offered by different types of institutes, whether it focuses on infrastructural support or purchase of technology equipment, and the scale at which the funding is provided.

5.2.2.2 The Global Environment Facility (GEF)

The GEF is the designated operating entity of the UNFCCC's Financial Mechanism.[5] Since its establishment in 1991, the GEF has allocated US$ 6.8 billion, which have been supplemented by more than US$ 36.1 billion in co-financing (GEF 2010a). These finances are raised through a process that is called the *GEF Replenishment*. Through this process, which consists of multi-year cycles, donor nations commit financial resources. Under its current Fifth Replenishment, the GEF Trust Fund has received a record replenishment of US$ 4.34 billion, of which approximately US$ 1.4 billion will be allocated to the GEF's Climate Change Focal Area Strategy during 2010–2014 (GEF 2011a).

Under the UNFCCC, in 2001, also the specific thematic funds Special Climate Change Fund (SCCF) and the Least Developed Country Fund (LDCF) were established (UNFCCC 2002). The SCCF funds projects related to:

(a) Adaptation;
(b) Technology transfer and capacity building;
(c) Energy, transport, industry, agriculture, forestry and waste management; and
(d) Economic diversification.

Among these four categories, adaptation is considered the top priority (GEF 2011b).

The LDCF is specifically designed to assist least developed countries in preparing and implementing national adaptation programmes of action (NAPAs). As per August 2009, the GEF mobilised US$ 180 million in voluntary contributions to the LDCF (GEF 2009). Since its inception in 1991, the GEF mobilised through the combined use of its *Strategic Priority for Adaptation*, the LDCF and the SCCF US$ 400 million for adaptation to climate change (GEF 2010a). Between July 2008 and June 2009, the GEF approved US$ 73 million in new investments in the LDCF and US$ 36 million of new investments in the SCCF (GEF 2010b). Additionally, under the *Poznan Strategic Programme on Technology Transfer* (UNFCCC 2008), the GEF allocated more than US$ 57 million that is supplemented by US$ 195 million in co-financing (GEF 2011a).

---

[5] In addition, the GEF is an operating entity of the Financial Mechanism of the Convention on Biological Diversity.

5.2.2.3 Multilateral, Bilateral and Regional Financing Institutes and Private Sector Investments

UNFCCC Art. 11 specifically refers to the possibility of developing countries having access to financial resources through bilateral, regional and other multilateral channels (UNFCCC 1992, Art. 11). In the meantime, a large variety of funding sources have become available through such channels. For example, the *Climate Investment Fund* (CIF) is a multilateral fund that channels finance through development banks (e.g., the African Development Bank, Asian Development Bank and the World Bank Group).[6] The CIF holds in total US$ 6.5 billion, of which US$ 4.5 billion is specified for the CIF's *Clean Technology Fund* and the remaining US$ 2 billion is specified for the *Strategic Climate Fund*. The Clean Technology Fund aims to demonstrate, deploy and transfer technologies for low-emission development, whereas the Strategic Climate Fund targets programmes to pilot new approaches to initiate transformation with potential for scaling up climate resilience (Adaptation Fund 2011).

The members of the *OECD's Development Assistance Committee* (DAC) have provided US$ 3.8 billion in bilateral Official Development Assistance (ODA) in 2007 to help developing countries reduce their own GHG emissions (about 4% of total bilateral ODA in 2007) (OECD 2011a). DAC data indicate that most support went into energy and transport sectors.

The record level of USD 630 billion in 2008 of Foreign Direct Investment (FDI) to developing countries marks the potential of private sector finance and investments to contribute to climate change mitigation and adaptation (UNCTAD 2010). For instance, UNCTAD estimated that low-emission FDI (for the renewables, recycling and low-carbon technology manufacturing sectors) into developing countries approached US$ 20 billion per year on average between 2003 and 2009 (UNCTAD 2010). Depending on the definition of low-carbon FDI, the OECD estimates that between US$ 7.6 and 190 billion has been invested through FDI flows to developing countries (OECD 2011a).

Other funds are available from, e.g., the World Bank Investment Framework for Clean Energy and Development, the UNDP Millennium Development Goal Carbon Facility, UNEP's Finance Initiative, and the European Commission Global Energy Efficiency and Renewable Energy Fund. According to UNEP, the estimated total of public climate financing reached nearly US$ 30 billion in 2009 (UNEP 2010). Of this, the multilateral development banks accounted for about USD 16.5 billion (UNEP 2010).

[6] Further information about the CIF can be found at http://www.climateinvestmentfunds.org/cif/. Accessed 28 September 2011.

5.2.2.4 Carbon Market Funding: The Clean Development Mechanism (CDM)

According to AGF (2010), the international carbon markets form a key source of finance for supporting mitigation and adaptation actions in developing countries. Currently, the main carbon markets internationally are the EU emissions trading scheme (ETS), which caps the GHG emissions of large emitters within the EU and enables these entities to trade emission allowances, and the CDM.

The CDM was introduced in 1997 in the Kyoto Protocol as one of the flexibility mechanisms to enable industrialised countries to achieve their quantified GHG emission reduction commitments partly through emission reductions projects in other countries. The CDM facilitates such project co-operation between industrialised and developing countries, whereby an industrialised country investor invests in a project in a developing country and receives GHG emission reduction credits (certified emission reductions or CERs) in return. A particular feature of the CDM is that the GHG emission reduction projects also need to contribute to sustainable development in developing countries.

According to AGF (2010), the CDM currently generates most of the funding on the carbon markets. For instance, from the EU, up to € 3 billion during the period 2013–2020 will be made available in the form of CDM projects (SEC 2011). In the 2005–2009 period, the CDM generated about € 5.8 billion in financial flows to developing countries (AGF Workstream 2011). As per mid-2011, more than 3000 CDM projects have been registered under the UNFCCC by the CDM's Executive Board (Fenhann 2011), which together would reduce just more than two billion tons of GHG emissions between 2008 and 2012 (Fenhann 2011). This makes the CDM the 'most significant financial flow in the global effort to address climate change' (AGF Workstream 2011).

Although the focus of CDM projects is clearly on mitigation (i.e., GHG emission reduction), in the design of the mechanism, a direct link has been established with adaptation. This has been done through the provision in Art. 12.8 of the Kyoto Protocol (UNFCCC 1998) 'that a share of the proceeds from certified [CDM] project activities is used... to assist developing country Parties that are particularly vulnerable to the adverse effects of climate change to meet the costs of adaptation'. Later during the negotiations, it was decided that 2% of the proceeds from monetised CERs is collected for supporting adaptation actions in developing countries. The money is collected and allocated by the Adaptation Fund under the Convention.[7] Up to July 2011, the Adaptation Fund had received more than 13 million CERs to its holding account,[8] which at the market prices around mid-2011 would represent a

[7] Further information about the Adaptation Fund can be found at http://unfccc.int/cooperation_and_support/financial_mechanism/adaptation_fund/items/3659.php Accessed 28 September 2011.

[8] http://cdm.unfccc.int/Issuance/SOPByProjectsTable.html. Accessed 30 June 2011.

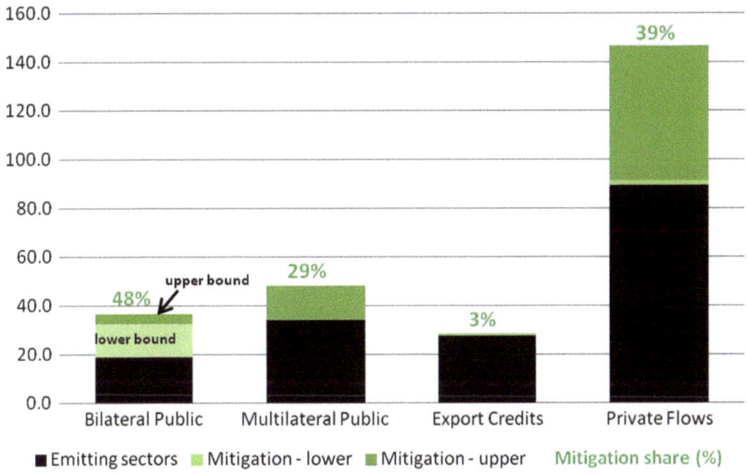

Fig. 5.2 Estimates of North-South financial flows, climate finance (mitigation) and climate-relevant (emitting) sectors (latest year estimates, billion USD). *Source* OECD (2011b)

value of more than US$ 160 million. As per 30 June 2011, the Adaptation Fund had allocated USD 60.6 million to ten adaptation projects and programmes (Adaptation Fund 2011). It is estimated that the Adaptation Fund will have approximately USD 334 million available by the end of 2012 (Adaptation Fund 2011).

There has been criticism of the CDM in terms of an unequal distribution of projects across developing countries. Especially, the share of least developed countries in the CDM pipeline has been relatively small. In these countries, the main potential for CDM projects exists in small-scale projects and for these projects, it is relatively difficult to cover project design, validation and verification costs from revenues from the CERs (Ellis 2006). A study by Bakker et al. (2009) has suggested several ways to modify the CDM structure so that stronger incentives can be created for projects in currently underrepresented regions, e.g., by 'overcrediting' projects from these regions.

To support these countries with developing CDM projects, there is now the possibility of formulating Programmes of Activities. These programmes can cover a range of similar small-scale activities so that instead of working through the full CDM project cycle with GHG accounting and accompanying costs for each activity in the programme, only the programme cycle as a whole will be assessed. This spreads the costs across a larger range of activities.

Finally, the ENTTRANS (2008) study on the potential of transferring and implementing sustainable energy technologies through the CDM argued that the CDM structure, with its strong focus on GHG abatement, has a tendency to lead to ad hoc investments without a clear link to the development strategies of host countries. Therefore, there is a risk that projects result in one-off investments without a strong spin-off potential into the country.

It can be concluded from the CDM experience that the project-based carbon market provides a large scope for public–private investor partnerships. Many CDM projects are financed through a collection of sources, such as investors participating in the project's equity, governmental support schemes, revenues from the project (e.g., electricity sold to the grid) and the (expected) value of the CERs. Such potential private–public partnerships will be discussed later in this chapter.

5.2.2.5 Discussion

The above overview has made clear that many activities have been initiated for financially supporting mitigation and adaptation measures. Figure 5.2 illustrates how the different categories described above (multilateral, bilateral, private sector, carbon market, etc.) have contributed to climate financing. Of these, the private sector currently provides the bulk of financial resources.

According to Practical Action (2010), a similar diversity can be seen in development assistance sources: Development finance sources are diverse and made available through governments, multilateral and bilateral agencies, civil society, the private sector and local finance initiatives. Through these sources, a variety of financing instruments can be used such as commercial funding, subsidies, concessional loans, grants and cross-subsidisation of end-user tariffs (Practical Action 2010).

On the one hand, this could enhance the overall climate finance system as the funds could complement each other (e.g., one fund focussing on infrastructure support and another source on actual technology capital investments) and spread the sources from different economic fields.

However, as Anderson (2011) explains, in spite of climate and sustainable development being interlinked, in practice it has turned out to be difficult for climate investment funds to simultaneously address socio-economic and mitigation and adaptation benefits: the so-called 'triple win'. He explains that while development agencies and banks are becoming increasingly interested in these multiple benefits, in particular 'the private sector cannot fully match the public investments needed to achieve the [Millennium Development Goals] and implement adaptation measures' (Anderson 2011, p. 3). In other words, finding synergies in combining different public and private finance mechanisms is a challenge in itself.

In addition, as is discussed later in this chapter, it is important that funding is not mainly focussed on projects reducing GHG emissions or enhancing adaptation, but that funding is provided through an integrated funding strategy whereby resource needs are identified for equipment purchase, operation and maintenance, as well as for creating an enabling environment for technology development and transfer, followed by an identification of possible private and public funding sources as described above. This will be discussed later in this chapter.

In the next section, we focus on the aspect that funding is often difficult to access by or is not affordable for large groups of people in developing countries, so that they cannot widen their economic opportunities, increase their energy options and diminish their vulnerability to external shocks (Practical Action 2010, p. 44).

5.3 Scaling Up Finance Initiatives for Realising Development Benefits

In an assessment such as TNA and LEDS, technologies and measures are prioritised for their contribution to achieving climate and development goals in developing countries. As the context for further development and transfer of these technologies is largely similar to that of development aid programmes, in this section, we discuss finance initiatives for realising development benefits such as energy access for poor population groups. From that experience, we can then draw conclusions on what this implies for identifying finance needs and resource allocation for low-emission and climate-resilient technologies for sustainable development identified in a TNA or LEDS.

5.3.1 Overview of Finance Patterns for Universal Energy Access

The IEA (2009) estimate that US$ 36 billion per year will be the funding required for enabling universal access to electricity. According to Practical Action (2010), this amount represents 3% of global energy investments while the target groups form around 30% of the world's population. Bazilian et al. (2011) include in their analysis the cost of universal access to modern energy services for cooking. In their estimate, the costs could even exceed US$ 70 billion per year on average to 2030. As mentioned earlier, these costs could become higher as climate change impacts could make it more difficult to achieve these energy access goals (Anderson 2011).

Bazilian et al. (2011) show that there is a variety of flows for financing global energy access. Based on an analysis of energy-related investments in Gross Fixed Capital Formation (GFCF),[9] FDI and ODA statistics, they conclude that:

1. Energy-related GFCF in developing countries has generally increased during the decade 2000–2009, but the diversity between countries is large with capital formation in Botswana increasing by 60% and in India by 160%, but decreasing in, for example, Niger.
2. No clear patterns can be found for electricity, gas and water-related FDI in developing countries as figures vary widely across countries.

[9] GFCF is the value of acquisitions of new or existing fixed assets by private and public sector entities.

3. There is an increasing trend in electricity-related ODA in non-OECD countries from US$ 5.6 billion in the year 2000 to US$ 9.67 billion in 2008.
4. These finance flows can be interlinked. For example, ODA could help improve the conditions in developing countries for attracting more FDI, which has been substantiated by UNCTAD (2010), p. 62).

Based on their assessment, Bazilian et al. (2011) conclude that existing country-specific investment flows in the energy sector are generally not sufficient to meet the needs of providing electricity services to those 1.4 billion people who currently lack any access. In some countries, average annual investment needed for universal household electrification is more than five times higher than energy-related GFCF. On average, Bazilian et al. (2011) estimate that for full household electrification in least developed countries, US$ 11.6 billion will be needed on average per year during 2011–2030. In 2008, actual investments on household electrification in least developed countries amounted to US$ 2.5 billion.

5.3.2 Need for System Improvements for Universal Energy Access

To enable funding to generate the maximum benefit for the resource, Bazilian et al. (2011) offer two important conclusions that are relevant for the challenge of scaling up financial initiatives for improved energy access:

1. For improving the investment climate for energy access, it is crucial that political commitments are secured and effective policy and regulatory, cross-sectoral frameworks put in place, so that existing market distortions can be removed and mechanisms made available to finance not only the equipment but also (pre-)feasibility studies, seed capital, insurance, debt and equity.
2. Finance flows are very diverse in terms of type, origin and destination and can vary from large infrastructural projects to small-scale purchases of biogas installations by farmers or improved cook stoves by households. When formulating strategies for technology innovation, as explained in Chap. 3, this diversity needs to be considered to identify the most suitable finance source for each action under a strategy.

A similar conclusion can be found in Practical Action (2010, p. 50) in their *Poor People's Energy Outlook*. They conclude that creation of markets should be an important aim of funding and that governments and donors have a critical role to play in removing the policy, regulatory and technical barriers that prevent small-scale finance from benefiting poor people:

> The goal is not simply to supply stoves and lamps or install generation capacity, but to provide the conditions for the creation of a sustained and profitable industry that will result in increased use of improved devices and generation capacity and will drive down costs (Practical Action 2010, p. 45).

Only if subsidies are well targeted to support market creation and governments and international organisations show strong leadership with pro-active policies and regulations, financing mechanisms such as microfinance, micro franchising, fee-for-service and carbon finance can bring benefits in terms of modern energy access in least developing countries at the desired scales.

Practical Action (2010) also explain how scaling up finance initiatives relates to the role of different stakeholders in a market, such as suppliers of the equipment, fuel providers, trainers and end-users. In addition, governments need finance for, e.g., building energy infrastructures and energy consumption in public buildings. It shows that funding energy access improvement or similar investments in energy efficiency and renewable energy sources in least developed, developing but also in developed countries is complex with a variety of items to be funded and diverse roles of different stakeholders. In the next part of this chapter, we discuss this issue in further detail with some examples of what project versus system- or market-level finance looks like.

5.3.3 Effectiveness of Financial Initiatives in Supporting the Poor

Effectiveness of aid is often expressed in terms of how it contributes to economic growth. However, such a macro-level focus does not necessarily reveal how effective aid is at the micro or project level. For example, Practical Action (2010, p. 44) concludes that particularly for people living in poverty, it is difficult to access the finance that they need:

> Many lack collateral and credibility, easy access to financial services, and the skills to convey entrepreneurship, fill in applications and make a business case. The marginalised poor are not integrated into the formal networks of society, which reduces funders' confidence in their ability to track their investments. Financial models to address the needs of poor people must focus on their common assets, such as resourcefulness and adaptability, strong informal social bonds with neighbours, family and community, entrepreneurial and opportunistic spirit, labour, and their willingness to invest the few resources they have for a better quality of life.

As is explained elsewhere in this chapter, recent trends in development aid have increased the focus to the creation of markets, so that the supply of equipment or other services is accompanied by the provision of conditions for sustainable industries. With these, the equipment can be better modified to local circumstances, maintained during its use and further improved nationally, thereby supporting further innovation processes. According to Practical Action (2010), the role of finance in this respect is dual. On the one hand, finance supports the growth of markets, but, on the other hand, finance should be organised in such a way 'that these markets are equitable for all participants'. Therefore, 'financing should aim to remove barriers, particularly those that affect marginalised people and to address gaps in support services, especially in poor or remote areas' (Practical Action 2010, p. 45).

The *Poor People's Energy Outlook* concludes that, generally, government-to-government and multilateral funding programmes pay relatively more attention to building up systems and markets in developing countries (Practical Action 2010, p. 46). Examples of this support are enhancing research and development in the country, training programmes, but also filling in gaps, such as facilitating subsidies, micro financing, fee-for-service and leasing-for-service for people who would otherwise not have access to private, commercial funding institutes (see Box 5.2). This addresses the point that system-level financing is usually more difficult to arrange through private sector financial means as expected financial returns are more difficult to predict than in case of a single project or programme of projects.

Furthermore, expected returns in a system-based finance approach are more heterogeneous and in some cases can provide a net loss to individual finance partners in the system. System-based finance is also more complex in that it requires co-ordinated actions of multiple (inter)national public and private finance providers at different levels of involvement/commitment, spanning a more heterogeneous scope of activities to be financed. The 'rules of the game' in a given country-context determine how financial risks and returns are divided between all financial stakeholders within a system-based approach.

Based on a survey by Wathne and Hedger (2010) among recipient countries for development assistance, multilateral organisations such as African Development Bank, World Bank, UNDP and Asian Development Bank were considered to contribute most strongly to, among others, the criterion of responsiveness to country circumstance and support for recipient-driven policies.

Private institutes provide funding on a commercial basis, and only those income groups who can afford the interest costs and repayments will have access to such funding. This explains why private institutes prefer financing projects over system improvements. The scope of the activities to be financed is relatively homogenous and can be clearly defined. These features ensure high levels of asset and activity controllability. Practical Action (2010), however, acknowledge that when projects are established at the level of a community, rather than for individual households, an investment and finance scale can be created that can attract private investors, especially when the investment is backed up by a governmental or multilateral organisation to support the improvements and thereby to lower the investment and default risk.

Box 5.2 Examples of instruments supporting people with poor access to finance

Some examples of instruments or tools to help people with poor access to finance acquire access to modern types of energy services are:

Subsidies Subsidies can be offered to support small-scale investments and can be in the form of financing partly the purchase of the equipment or the energy resources needed for the equipment. In case of the first, subsidies would help reduce the up-front costs of using a technology. Subsidising purchase of fuels without subsidising the purchase of the equipment may reduce fuel costs but may lead to situations in which people purchase equipment which will be cheaper to use, but for which the investment costs may turn out to be unaffordable.

Microfinance This mainly consists of small loans to people for whom the commercial finance market contain too many barriers for obtaining loans. Microfinance schemes have lower administrative costs and could arrange that groups of borrowers assist each other in paying back loans. Interest rates could be relatively high though, as the clients often have no assets to offer as collateral. Also, borrowed money could be partly used for personal matters because of unsustainable business management, so that additional loans are needed. This may, for instance, place women, who take microcredits, in a vulnerable position when they have difficulties in paying the loan back and are blamed by their families for getting them into debt.

Fee-for-service In this model, customers can purchase equipment without the need for buying ancillary devices, as the use of the latter can be offered by nearby service points. For example, people could purchase a lantern or other device and charge it at a local charging station. By doing so, purchase of their own charging equipment is no longer needed; people only have to pay a fee for the charging service (Practical Action 2010). To make this service profitable, there has to be sufficient customers to use the service stations.

Leasing for service Instead of purchasing the equipment, it could also be leased (possibly in combination with fee-for-service) so that relatively high purchase costs could be avoided. On the other hand, leasing requires payment of a regular fee, so that in the longer run, people may be worse off unless the leasing company also periodically renews the equipment being leased.

Micro franchising This concept aims at providing business opportunities and services in poor regions in developing countries by creating a brand and then identifying community entrepreneurs. These entrepreneurs are then trained in business management techniques with the help of proven business models, both initially and while undertaking the business. The objective is to improve business knowledge in poor regions so that failure rates of businesses are reduced.

5.3.4 Overall Lessons from Aid Effectiveness

The complexity of development aid has resulted in numerous investigations into how development finance can be most effectively allocated. In 2005, the OECD DAC formulated the lessons learned from these investigations in a new agreement: the *Paris Declaration on Aid Effectiveness*. This Declaration states a number of prerequisite principles for effective aid (Kerkhoff et al. 2011, OECD 2005/2008):

- *Ownership*: Developing countries must lead their own development policies and strategies and manage their own development work on the ground. Capacity to exercise this kind of leadership by strengthening local expertise, institutions and management systems needs to be supported.
- *Alignment*: Donor countries must line up their aid firmly behind the priorities outlined in developing countries' national development strategies.
- *Harmonisation*: Actions of donors need to be harmonised to avoid duplication and high transaction costs for developing countries. Countries agreed on a target that two-thirds of all aid should be provided via programmes in which funding is pooled to support a particular recipient country's strategy, instead of funding multiple individual projects.
- *Managing for results*: Aid should have a stronger focus on results so that it becomes clear what tangible difference it makes for improving the livelihood of the poor.
- *Mutual accountability between donors and partner countries*: Donors and recipient developing countries must account more transparently to each other how their aid funds are used, and what impact aid has had.

At the *Accra Forum on Aid Effectiveness* of 2008 (OECD 2005/2008), the principles agreed at Paris in 2005 were largely reinforced. However, it emphasised the increasingly important paradigm of aid as a partnership between several players, such as global funds, countries, private sector and local organisations. Through partnerships, different stakeholders would be able to work with similar principles and procedures so that efforts could become more coherent.

A clear example of what such partnerships could look like, focussing on environmentally sustainable energy technologies in developing countries, is provided by the *Renewable Energy and Energy Efficiency Partnership* (REEEP), which brings together several public and private sector donors for financing projects and strategy support programmes in developing countries (see Box 5.3).

Box 5.3 The Renewable Energy and Energy Efficiency Partnership (REEEP)
REEEP is a non-profit specialist change agent that aims at catalysing markets for renewable energy and energy efficiency. Its primary focus is on emerging markets and developing countries. It was established alongside the 2002 World Summit on Sustainable Development in Johannesburg, South Africa. The partnership initiates and funds projects to support the following areas for sustainable energy:

- Assisting governments in creating favourable regulatory and policy frameworks.
- Promoting innovative finance and business models to activate the private sector.

REEEP develops tools and programmes for developing country energy sector practitioners to acquire information about renewable energy and

energy efficiency technologies and how to technically and financially support these. REEEP is now composed of more than 350 partners, including 45 governments as well as a range of private companies and international organisations.
Through its regional secretariats, including China and India, activities can be kept locally relevant and focused.
REEEP is supported primarily by governments (Australia, Austria, Canada, the EU, Germany, Ireland, Italy, the Netherlands, New Zealand, Norway, Spain, USA and UK) and by contributions from the private sector. Robust governance and monitoring structures ensure that all contributions are spent transparently and effectively.
Source http://www.reeep.org. Accessed 20 September 2011.

Finally, the Network on Environment and Development Co-operation (ENVIRONET) is an example of a global network (managed by the OECD DAC) to promote and facilitate the integration of environment and climate change into all aspects of development assistance.[10]

5.3.5 Discussion on Implications for Accelerating Low-Emission Strategies

The experience with development aid, as discussed earlier, has made clear that the effectiveness of financial initiatives is generally enhanced if the funding is focussed on an overarching strategy instead of individual projects, with active government support and commitment. In such a strategy, an investment or a series of investments in technologies or measures are supported by infrastructural improvements, supporting services and streamlined implementation chains of market actors. Depending on the scale of individual investments and the affordability of financial products, choices can be made between, e.g., capital investments, debt finance and acquiring public funding, as well as whether to purchase equipment, in combination with, e.g., fee-for-service or lease it.[11]

The implication of these insights from development aid effectiveness for the financing of strategies and action plans from a TNA or LEDS is that a funding strategy for these outputs needs to focus on identifying what needs to be funded and what is the most suitable finance source.

[10] http://www.oecd.org/dac/environment. Accessed 27 September 2011.

[11] At the project level, UNFCCC (2006) supports and guides developing countries in preparing technology transfer projects for financing. In addition, the CTI *Private Financing Advisory Network* (PFAN) offers supporting services for financing to developing countries (http://www.cti-pfan.net/, accessed 18 September 2011).

The complexity of formulating a funding strategy as part of a country's technology strategy is that the technologies can have different scales of implementation and could be at different stages of development. Scale and availability in time can have implications for finance risks and this has consequences for the availability of, e.g., commercial financial resources and affordability of finance for end users. For completion of a funding strategy for TNA and LEDS outputs, therefore, a clearer insight is needed on the interplay between private and public finance institutions and initiatives, and this will be discussed in the next section. Analysis of aid effectiveness is relevant for this as the development and transfer of low-emission and climate-resilient technologies identified for sustainable development are likely to take place in comparable markets and systems as has been the case with development aid.

5.4 Towards Integrating Funding Strategies for Climate and Development

5.4.1 Funding Needs for an Innovation Strategy

Thus far in this chapter, we have explored finance needs for achieving climate and development goals. For the mitigation and adaptation part, we generated an overall idea of finance needs, based on larger-scale economic analysis, and how much funding has already been made available by various sources (sub-challenge 1 in this chapter). We also explored possible ways for scaling up finance actions for achieving development goals, such as improved energy access in developing countries (sub-challenge 2).

In this section, we focus on sub-challenges 3 and 4 in this chapter by exploring how a country could formulate, based on the outputs of a TNA or LEDS (discussed in Chaps. 2 and 3), *an integrated funding and innovation strategy* for prioritised technologies and measures, including identification of financial sources for the different components of the strategy

As explained in Chap. 2, TNA and LEDS result in portfolios of prioritised technologies and measures for climate and development with estimates of overall benefits and an overall cost figure in terms of capital investment and operational and maintenance costs at the desired scale for a particular period of time (e.g., 20 years, depending on whether the technology is available in the short or long term).

As a next step, actions for accelerating the technology development, deployment or diffusion are identified and prioritised as input in a strategy (Chap. 3). The actions are prioritised and each priority action is characterised in an Action Plan in terms of, among others, timing, responsibility and costs, such as how much does it cost to set up a training programme for domestic engineers to work with this technology?

Table 5.3 Overview of funding components for an integrated strategy for TNA & LEDS outputs

TNA & LEDS stage	Item to be financed	Cost (US$)
Prioritised technologies and measures in TNA & LEDS (Chap. 2)	Purchase of equipment and operation and maintenance (over selected lifetime)	US$ (see Chap. 2, Table 2.1)
Prioritised actions for technology or measure development and transfer (Chap. 3)	Funding of actions for improvement of enabling environment for technology or measure	US$ (see Chap. 3, Table 3.6)
Total finance need for acceleration of TNA & LEDS output in the country		US$ equipment + US$ enabling environment

Table 5.3 summarises these two main TNA and LEDS components of the finance needs for the acceleration of development and transfer of prioritised technologies in a country:

1. The purchase of the technology equipment or implementation of mitigation and adaptation measure.
2. The supporting actions for technology acceleration within a country, which is mainly related to system-level improvements: Finance needs per (sub)sector are obtained by aggregation and rationalisation of actions across technologies within the (sub)sector, whereas aggregation across (sub)sectors results in funding needs for a national strategy.

The next question is how these finance needs can best be met and how to address the financial risks to generate a plan for funding the whole innovation strategy. To do this, it is important to obtain a clearer insight in the interplay between private and public sector financial sources in financing equipment purchase and creating an enabling environment for technology transfer.

5.4.2 Interplay Between Private and Public Funds for Technology and Innovation Support Financing

In Sect. 5.3, it has been explained that, based on experience with energy access improvement work in developing countries, system improvements in developing countries tend to be undertaken by bilateral and multilateral funders. Examples of such investments are 'the creation of research institutions and laboratories, microfinance institutions for the poorest and for reducing risk to the banking sector and unfavourable policies or policy gaps' (Practical Action 2010, p. 45).

Private institutions, on the other hand, provide funding on a commercial basis, and only those income groups who can afford the interest costs and repayments will have access to such funding.

Translating this to the financing of the outputs of TNA and LEDS, it would imply that private sector funding could be mainly attracted for technology equipment that is available in the short term at commercial terms. However, at the UNFCCC TNA workshop in Bonn (1–2 June, 2011), several international funding organisations emphasised that while commercially viable technologies are relatively easy to deploy, projects will not be developed if the relevant policy frameworks are absent (ENB 2011).

With these two general observations in mind, we can have a further look at how private and public finance initiatives can jointly support TNA and LEDS technologies and strategies for low-emission and climate-resilient development.

According to Nassiry and Wheeler (2011), private investment in low-emission technologies is limited by a number of aspects. First, investment in low-emission technologies that are still in RD&D stage and, therefore, available only in the longer term is 'hindered by longer investment periods before exit, more capital-intensive development that requires large follow-on financing, smaller investment sizes coupled with similar due diligence costs and management fees, and higher execution risks than later-stage financing'.

In their view, this even hinders the mobilisation of venture capital, which can usually play a critical role in the early and growth stages of low-emission and climate-resilient technology investment.

These observations would lead to the conclusion that private sector capital, including venture capital, private equity, infrastructure fund investments[12] and debt finance through financial markets could mainly be appropriate for technologies that are in a later stage of deployment and diffusion.

To illustrate this, Nassiry and Wheeler (2011) estimate that global investment in low-emission energy production reached US$ 243 billion in 2010 (an increase from the 2009 level of 30%), but that only US$ 2.1 billion was attracted through venture capital for early stage technology development and US$ 6.6 for later stage venture capital and private equity.

In other words, based on present private capital investment flows in technology development and transfer, it can be concluded that their main focus is on technologies that are in a later stage of development.

A similar point has been made by the concept of *Valley of Death*, as developed by the World Bank (see Fig. 5.3). It explains how public development efforts play a key role during the initial technology development stage with a more important role for the private sector during commercial application of the technology.

A crucial phase is in between of these two stages, which is called the 'Valley of Death'. Here, public development efforts should be replaced with private development efforts, but this can be hampered because of risk, unfamiliarity with the

[12] Venture capital is equity capital with a focus on relatively early stage technologies; investors are willing to accept higher risks for a relatively high internal rate of return (50% or higher). Private equity has a stronger focus on more mature technologies with medium risk and internal rates of return of around 25%. Infrastructure funds focus mainly on infrastructural investments with long duration, steady, low risk and internal rates of return of around 15% (Nassiry and Wheeler 2011).

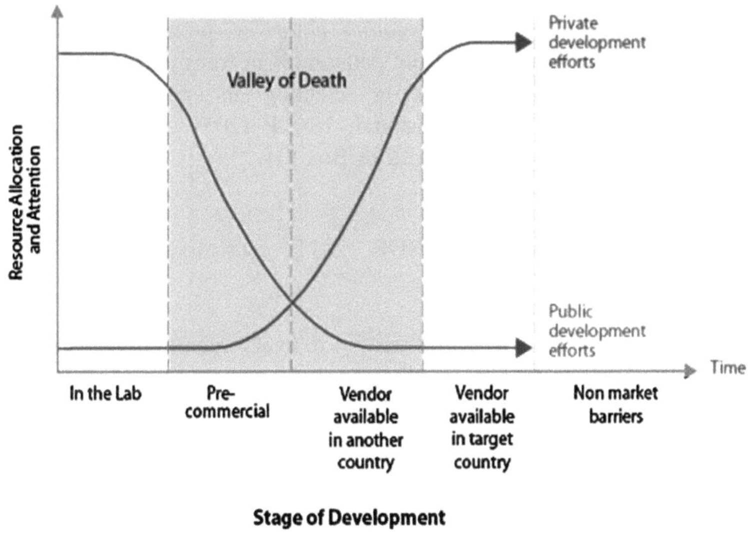

Fig. 5.3 Valley of Death concept illustrated (Jonathan Coony, World Bank Group) (UNFCCC 2011b, UNDP 2010)

technology, inefficient enabling environment, etc. In other words, if decreasing public sector support is not replaced with private development efforts, then the technology development could be stopped here (UNDP 2010).

For technologies in the stage of RD&D, public finance is the main source of support. Nassiry and Wheeler (2011, p. 3) summarise this as follows:

> Public subsidies will be needed to accelerate innovation and investment, promote learning and scale economies, and progressively reduce costs to the point where commercialisation and deployment of low-carbon technologies become attractive to more private investors.

To support this process, Nassiry and Wheeler (2011) suggest *public–private green venture funds* in which public funding is used as a leverage for attracting private funding for both technology development and deployment. In their proposal, the fund would be a fund of funds, whereby different private and public sources are collected into one fund that undertakes the investments.

In a similar vein, but then more with a focus on private–public partnership for *debt financing* for low-emission and climate-resilient technologies, UNDP (2011) explains the example of public-guaranteed *green bonds* in combination with commercial bank loans as a tool to support technology investments. UNDP (2011) describe how countries such as Brazil, China and India have gained experience with commercial bank finance underwritten by government guarantees for low-emission technologies.

Finally, UNDP (2011) gives the example of the *Global Energy Efficiency and Renewable Energy Fund (GEEREF)*, launched by the European Commission in

2008, for providing public–private *equity* aimed at promoting clean energy investment in developing countries and economies in transition.[13]

To support developing countries in selecting an appropriate policy and financing mix to catalyse climate capital, UNDP (2011, Chap. 5) presents a stepwise approach, which is reproduced in Box 5.4.

Box 5.4 Steps identified in UNDP (2011) catalysing climate finance guidebook

- Step 1. Assess the financing terms and delivery requirements (financing scale, disbursement timing, etc.) of the envisaged public policy measures to support technology development and deployment.
- Step 2. Take into account the constraints of all parties concerned, and notably private investors in public–private partnerships, prioritize possible uses of public funds.
- Step 3. Review country and thematic eligibility criteria to existing relevant international and national public funds.
- Step 4. Identify possible gaps and overlaps in funding sources.
- Step 5. Identify possible cross-sectoral blending of finance to address gaps and minimise overlaps.
- Step 6. If necessary, adjust the mix/sequencing of public policy instruments to address financial constraints.

How the actions identified for technology acceleration strategies (as discussed in Chap. 3) could be financed by private and public funds, or a combination of these, among others, depends on the characteristics of the actions, just as technology characteristics determine what would be appropriate funding.

For instance, as described earlier, private sector infrastructure funds could support improving the enabling environment for technology roll-out at the desired scales. When for a technology roll-out a training programme is considered necessary, then this could be a part of venture capital fund participation as the training would improve the performance of the overall investment strategy. When the action would involve improving the system for energy services, for instance, build-own-transfer (BOT) financing could be applied, such as done in some large infrastructure projects, through investments and operation by private investors. After a specific period of time, the ownership of the investment can be transferred to public owners.

[13] http://geeref.com. Accessed 28 September 2011.

However, for other system improvements, such as implementation and enforcement of new policies and measures or organisational change or market support such as quality control, etc., public funds will be required.

Given the current flows of private and public finance to development and climate investments in developing countries, it seems reasonable to conclude that:

- Investments in actions to create an enabling environment within a country for accelerating technology development and transfer at the desired scale for climate and development goals have not so far been fully accounted for and mainly will have to come from public sector funds (including bilateral and multilateral funds).
- Public sector funding also has to take the bulk of investments in technologies that are still in the early stages of innovation (RD&D).
- Public–private funding partnerships, e.g., through venture or equity capital, could support deployment of later-stage technologies.
- Mainly later-stage technologies (near or at commercial availability), backed up by stable policies, infrastructures and possibly underwritten by government guarantees, are attractive for commercial debt finance.

5.4.3 An Integrated Finance and Innovation Strategy

5.4.3.1 Integrating a Finance and Innovation Strategy for Coherent Implementation

In the preceding discussion, we concluded that private financial investments in low-emission and climate-resilient technologies are largely focused on technologies that are at a later stage of development (i.e., closer to commercial application) and that the initial RD&D stages largely depend on public finance initiatives. We have seen examples, however, of combined private–public finance initiatives where public funding is used to mitigate investment risks and/or create enabling environments for technologies to develop further.

With these insights, it is suggested that it would be useful if country stakeholder groups in TNA and LEDS, with support from the TEC and CTCN, could formulate a funding strategy that could be integrated with a technology acceleration strategy for a sector or country.

For the funding strategy, stakeholders can identify what components in a technology strategy (equipment and technology acceleration actions) need to be funded (see Table 5.3) and allocate these funding needs across public and private finance initiatives. For example, for a prioritised technology that is in the process of deployment in the market, a combination of venture capital, private equity and debt finance can be considered, whereas public and private infrastructural funds can be approached for market or system improvements. For funding of training programmes, a multilateral fund or bilateral programme can be approached. The

MRV activities for the programme's performance could, for instance, be funded by the country's government itself. The example is purely hypothetical and several combinations are possible for an optimal funding mix between public and private finance initiatives and products within the specific technology transfer context.

The integration of a technology and funding strategy is desirable as for successful transfer and innovation of technology into a country system all the actions identified in the action plan and later on in the rest of the strategy are required. In addition, treating the funding of technology and innovation acceleration actions as an integrated whole over time maximises the chances of success and avoids funding in a piecemeal fashion.

In line with the conclusions in Chaps. 3 and 4, it is important that financial needs and resources for a technology-, sector- or national-level strategy are rationalised to avoid duplications in funding or gaps. Just as technology transfer at the desired scale requires a strategy with rationalisation of acceleration actions, financing technologies and enabling environment improvements require a rationalised and integrated funding strategy for the specific country innovation context.

5.4.3.2 Structural Integration of Financial Elements into the Technology Transfer Mechanism

The structural aspects of the Technology Mechanism were discussed in Chap. 4. It was argued that for coherent support for delivery of a TNA and LEDS, the elements of adaptation and capacity building had to be fully integrated with technology transfer under the Technology Mechanism, rather than dealt with separately within the different pillars under the convention. From the discussion on finance, it would also be beneficial if finance support to developing countries on appropriate combinations of private and public funding instruments could also come under the Technology Mechanism to provide a funding strategy so that implementation of action plans for technologies could be based on fully rationalised and coherent actions and funding plans.

This would help to ensure consistency of implementation over time within a country for allocations and for MRV activities. Another benefit would be the ability to aggregate the information across countries and regions for development of international supporting policies.

5.5 Meeting the Challenge: Integrated Funding Strategies for Climate and Development Innovation

To realise developing country action, financial support is crucial. Until now, the estimations of required funding for mitigation and adaptation have concentrated mainly on costs of the technology or measure. Therefore, the scale of funding required for mitigation and adaptation is likely to be much higher as also costs associated with the actions needed for technology-, sector- and national-level strategies will have to be covered.

Similarly, it can be concluded that although the international 'green finance industry' has substantial funds at its disposal to support developing countries in their transition to a low-emission sustainable future (including the finances pledged and allocated under the Cancun Agreements), it seems likely that additional funding will be needed to finance necessary improvements of the enabling environment for technology development and transfer. In particular, with respect to the latter aspect, scaling up public finance initiatives, possibly in combination with private initiatives, will be required.

In line with that, we concluded that development banks increasingly focus on national development goals, not just technology dissemination through projects (ENB 2011). Similarly, the European Union, building on the AGF (2010) report, concludes that for 'maximum coherence between development aid and climate finance, the most important approach in each country or region will be to integrate development and climate change challenges in one single low-carbon development strategy' (SEC 2011). However, although there is evidence that international financial resource providers value highly the alignment of project proposals with the country's national sustainable development objectives to maximise benefits, in practice they find this difficult to do.

The TNA and LEDS processes would help achieve these aims especially if attention is paid to developing strategies for enabling energy access for the poor that may involve new financial instruments or combinations and has to involve development NGOs in the early stages.

The evidence from aid effectiveness and from innovation studies has also shown that finance initiatives will eventually be most effective if technology investments are combined with system improvements. This is what the country processes for TNA and LEDS provide: They generate costs both for technologies and measures and for innovation system improvements within action plans (accompanied by MRV and with timescales). This facilitates resource allocations for funding.

Different funding providers would be expected to be involved in funding different activities under the strategies developed. Private sources of funding can be expected to cover a portion of the mitigation and adaptation costs in sectors with privately owned physical assets. However, public resources will be needed to implement the innovation support actions, including policies or regulations to encourage the investment of private resources in mitigation or adaptation measures, especially in developing countries. Public domestic resources will also be needed to cover adaptation costs related to climate change impacts on public infrastructure (UNFCCC 2008).

With the large number of private and public players identified earlier and the need to have both technologies and measures and the innovation strategy capacity-building actions financed, we recommend that a finance strategy is generated and integrated with the TNA and LEDS innovation strategy to maximise successful transfers. Support for this process from the Financial Mechanism under the Convention will be very important as developing countries differ substantially in their capacity to access these financial resources.

In Chap. 4, we proposed that technology transfer, adaptation, mitigation and capacity building are integrated for efficient delivery of the TNA and LEDS outputs. In this chapter, we now propose that for efficient implementation of the action plans, we also need a further integration of finance support activities under the Technology Mechanism rather than dispersed across adaptation, capacity building and finance entities. In a resource-constrained world, any funding, therefore, has to be efficiently directed to address both climate goals for mitigation and adaptation and sustainable development objectives.

In addition, the streamlining proposed for the financial funds under the Convention to be devised by the new Standing Committee should take this into account in rationalising and improving the operation of the range of financial entities.

Thus, the sub-challenges outlined at the start of the chapter have each been addressed with exploration of some of the wider implications for overall structures and for funding scales.

References

Action Aid (2009). *Equitable adaptation finance: The case for an enhanced funding mechanism under the UN framework convention on climate change*. Retrieved September 28, 2011 from http://actionaidusa.org/assets/pdfs/climate_change/equitable_adaptation_finance.pdf

Adaptation Fund (2011). *TC Workshop on lessons learned from relevant funds and institutions for the design of the GCF: The Adaptation Fund Experience*. 12 July 2011. Retrieved September 26, 2011 from http://unfccc.int/files/cancun_agreements/green_climate_fund/application/pdf/tc2_workshopadaptationfund.pdf

AGF (2010). *Report of the Secretary General's High Level Advisory Group on Climate Change Financing*. UN High-level Advisory Group on Climate Financing. 5 November 2010. Retrieved September 28, 2011 from http://www.un.org/wcm/webdav/site/climatechange/shared/Documents/AGF_reports/AGF%20Report.pdf

AGF Workstream 8 (2011). *Carbon markets: Benefits to developing countries and options for expansion*. UN High-level Advisory Group on Climate Financing. Retrieved September 28, 2011 from http://www.un.org/wcm/webdav/site/climatechange/shared/Documents/AGF_reports/Work_Stream_8_%20Carbon%20markets.pdf

Anderson, S. (2011). *Climate change and poverty reduction*. Climate & Development Knowledge Network (CDKN). Policy Brief, August 2011. Retrieved August 9, 2011 from http://cdkn.org/wp-content/uploads/2011/08/CDKN_poverty-reduction_FINAL.pdf

Bakker, S. J. A., Asselt, H. D. van, Gupta J., Haug, C., & Mar, S. (2009). *Differentiation in the CDM: Options and impacts*. Report 500102 023. ECN-B-09-009. Amsterdam, The Netherlands. Retrieved September 28, 2011 from http://www.pbl.nl/bibliotheek/rapporten/500102023.pdf

Bazilian, M., Nussbauer, P., Gualberti, G., Haites, E., Levi, M., Siegel, J., Kammen, D.M., & Fenhann, J. (2011). *Informing the financing of universal energy access: An assessment of current flows*. Nota di Lavoro. Fondazione Eni Enrico Mattei. 56.2011. Retrieved September 24, 2011 from http://www.feem.it/userfiles/attach/20117181213234NDL2011-056.pdf

Ellis, J. (2006). *Issues related to a programme of activities under the CDM*. OECD, COM/ENV/EPOC/IEA/SLT (2006)3. Retrieved September 28, 2011 from http://www.iea.org/papers/2006/CDMissues.pdf

ENB (2011). Summary of the UNFCCC workshop on technology needs assessments: 1–2 June 2011. *Earth Negotiations Bulletin, 12*(501). Retrieved September 27, 2011 from http://www.iisd.ca/download/pdf/enb12501e.pdf

ENTTRANS (2008). *Promoting sustainable energy technology transfers through the CDM: Converting from a theoretical concept to practical action*. European Union Sixth Framework Programme. Retrieved September 23, 2011 from http://jiqweb.org/images/stories/mifiles/downloads/jin/ENTTRANSd2.pdf

Fenhann, J. (2011). *UNEP Risoe CDM/JI pipeline analysis and database*. Retrieved September 1, 2011 from http://cdmpipeline.org/

GEF (2009). *LDCF fact sheet*. Retrieved September 28, 2011 from http://www.thegef.org/gef/sites/thegef.org/files/publication-/LDCF-factsheets09_0.pdf

GEF (2010a). *Financing adaptation action*. Retrieved September 28, 2011 from http://www.thegef.org/gef/sites/-thegef.org/files/publication/adaptationactions_0.pdf

GEF (2010b). *GEF annual report 2009*. Retrieved September 28, 2011 from http://www.thegef.org/gef/pubs/annual_report_2009

GEF (2011a). *Implementing the Poznan Strategic Program on Technology Transfer*. Retrieved September 28, 2011 from http://www.thegef.org/gef/sites/thegef.org/files/publication/Tech Transfer_2010.pdf

GEF (2011b). *Strategy on adaptation to climate change for the Least Developed Countries Fund (LDCF) and the Special Climate Change Fund (SCCF)*. Retrieved September 28, 2011 from http://www.thegef.org/gef/sites/thegef.org/files/publication/GEF-ADAPTION%20STRATEGIES.pdf

IEA (2009). *World Energy Outlook 2008*. International Energy Agency. OECD/IEA, Paris. Retrieved May 28, 2009 from http://www.iea.org/weo/

IEA (2010). *Energy technology perspectives 2010: Scenarios and strategies to 2050*. Paris: OECD/International Energy Agency.

IPCC (2007). Climate change 2007: Synthesis report. In Pachauri, R. K & Reisinger, A. (Eds.) *Contribution of Working Groups I, II and III to the Fourth Assessment Report of the Intergovernmental Panel on Climate Change*. Geneva: IPCC.

Kerkhoff, L. van, Habib Ahmad, I., Pittock, J., & Steffen, L. (2011). *Designing the Green Climate Fund: How to spend $100 billion sensibly*. Retrieved September 17, 2011 from http://www.environmentmagazine.org/Archives/Back%20Issues/2011/May-June%202011/green-climate-fund-full.html

Nassiry, D., & Wheeler, D. (2011). *A green venture fund to finance clean technology for developing countries*. Working paper 245, Center for Global Development. Retrieved September 28, 2011 from http://www.cgdev.org/files/1424899_file_Nassiry_Wheeler_Green_Venture_Fund_FINAL.pdf

OECD (2005/2008). *The Paris declaration on aid effectiveness and the Accra agenda for action*. Retrieved September 28, 2011 from http://www.oecd.org/dataoecd/11/41/34428351.pdf

OECD (2011a). *Defining and measuring green FDI: An exploratory review of existing work and evidence*. OECD Working Papers on International Investment. No. 2011/2. Retrieved September 28, 2011 from http://www.oecd.org/dataoecd/26/8/48171454.pdf

OECD (2011b). *Financing Climate Change. Better Policies for Better Lives*. Organisation for Economic Cooperation and Development. Paris, France. Retrieved December 20, 2011 from http://www.oecd.org/env/cc/financing

Practical Action (2010). *Poor people's energy outlook 2010*. Rugby, UK. Retrieved August 15, 2011 from http://practicalaction.org/ppeo2010

Project Catalyst (2010). *From climate finance to financing green growth*. Briefing paper, 23 November 2010. ClimateWorks Foundation and the European Climate Foundation. Retrieved September 28, 2011 from http://www.projectcatalyst.info/images/publications/101127_from_climate_finance_to_financing_green_growth_formated.pdf

SEC (2011). *Commission staff working document: scaling up international finance after 2012*. European Commission. Retrieved September 28, 2011 from http://ec.europa.eu/economy_finance/articles/financial_operations-/pdf/sec_2011_487_final_en.pdf

Stern, N. (2007). *The economics of climate change: The Stern Review*. Cambridge: Cambridge University Press.

UNCTAD (2010). *World Investment Report 2010: Investing in a low-carbon economy*. United Nations Conference on Trade and Development. Retrieved September 28, 2011 from http://www.unctad.org/en/docs/wir2010_en.pdf

UNDP (2007). *Development Report 2007/2008*. New York, New York. Retrieved January 28, 2011 from http://hdr.undp.org/en/media/HDR_20072008_EN_Complete.pdf

UNDP (2010). *Handbook for conducting technology needs assessment for climate change*. Retrieved September 23, 2011 from http://unfccc.int/ttclear/pdf/TNA%20HANDBOOK%20EN%2020101115.pdf

UNDP (2011). *Catalysing climate finance: A guidebook on policy and financing options to support green, low-emission and climate-resilient development*. New York, NY. Retrieved September 28, 2011 from http://content.undp.org/go/cms-service/download/publication/?version=live&id=3267712

UNEP (2010). *Bilateral finance institutions and climate change: A mapping of 2009 climate financial flows to developing countries*. Retrieved September 28, 2011 from http://www.unep.org/pdf/dtie/BilateralFinanceInstitutionsCC.pdf

UNFCCC (1992). *United Nations Framework Convention on Climate Change*, FCCC/INFORMAL/84. Retrieved May 23, 2011 from http://unfccc.int/resource/docs/convkp/conveng.pdf

UNFCCC (1998). *Kyoto Protocol to the United Nations Framework Convention on Climate Change*. Retrieved May 23, 2011 from http://unfccc.int/resource/docs/convkp/kpeng.pdf

UNFCCC (2002). *Report of the Conference of the Parties on its seventh session: Addendum part two: Actions taken by conference of the parties*, FCCC/CP/2001/13/Add.1, Decision 7/CP.7. Retrieved September 28, 2011 from http://unfccc.int/resource/docs/cop7/13a01.pdf#page=22

UNFCCC (2006). *Preparing and presenting proposals: A guidebook on preparing technology transfer projects for financing*. Issued by the Climate Change Secretariat (UNFCCC) Bonn, Germany. Retrieved September 27, 2011 from http://unfccc.int/ttclear/pdf/PG/EN/UNFCCC_guidebook.pdf

UNFCCC (2007). *Investment and financial flows to address climate change*. Retrieved September 28, 2011 from http://unfccc.int/files/cooperation_and_support/financial_mechanism/application/pdf/background_paper.pdf

UNFCCC (2008). *Development and transfer of technologies*. Decision 2/CP.14, FCCC/CP/2008/7/Add.1. Retrieved September 21, 2011 from http://unfccc.int/resource/docs/2008/cop14/eng/07a01.pdf#page=3

UNFCCC (2010). *The Cancun Agreements: Outcome of the work of the Ad Hoc Working Group on Long-term Cooperative Action under the Convention*. Decision 1/CP.16. FCCC/CP/2010/7/Add.1. Retrieved July 17, 2011 from http://unfccc.int/resource/docs/2010/cop16/eng/07a01.pdf#page=2

UNFCCC (2011a). *Synthesis report on the national economic, environment and development study (NEEDS) for climate change project*. FCCC/SBI/2010/INF. Retrieved September 28, 2011 from http://unfccc.int/resource/docs/2010/sbi/eng/inf07.pdf

UNFCCC (2011b). *Enhancing the implementation of the results of TNAs*. Background Paper II for UNFCCC workshop on technology needs assessments, Bonn, Germany, 1–2 June 2011.

Retrieved June 3, 2011 from http://unfccc.int/ttclear/pdf/Workshops/Bonn%202011/BG%20paper%20II-implementing%20TNAs.pdf

Wathne, C., & Hedger, E. (2010). *What does an effective multilateral donor look Like?* Project Briefing, No. 40. Overseas Development Institute. Retrieved September 16, 2011 from http://www.odi.org.uk/resources/download/4800.pdf

WRI (2011). *Summary of developed country 'fast-start' climate finance pledge*. World Resources Institute. Retrieved September 29, 2011 from http://www.wri.org/publication/summary-of-developed-country-fast-start-climate-finance-pledges

Chapter 6
Summary of the Challenges and Solutions

Abstract The important challenges are listed below and have formed the subject of each of the chapters in the book. This chapter summarises the key insights gained from the earlier chapters and indicates some possible ways forward. There are no simple solutions and real progress will involve the political will commitment to change, many stakeholders and networks working flexibly together as well as sustained effort and long term finance.

6.1 Introduction

We began this book with an overview of recent insights in the risks associated with allowing greenhouse gas (GHG) emissions to increase along business-as-usual pathways. The conclusion was crystal clear. Although there has been decreasing public and political interest in climate change during the past 3 or 4 years, the problem will not solve itself. Not only do we run the risk of future climatic impacts to ecosystems, we can already see changes to our environment and increased frequency of extreme weather events. This shows that climate change poses increasing threats to our security, to our food supplies and to the availability of water and land as well as threats to the ecosystems on which we depend for oxygen production, food production and water circulation.

At the same time, certain mineral resources are becoming scarce, and price volatility for commodities is causing hardship round the planet with developing countries paying the highest price. For example, in this book we discussed that for African countries, the costs of reaching the Millennium Development Goals could become 40% higher than expected because of the need to adapt to climatic changes (Anderson 2011).

W. van der Gaast and K. Begg, *Challenges and Solutions for Climate Change*, Green Energy and Technology, DOI: 10.1007/978-1-84996-399-2_6,

Therefore, to continue business as usual is only in the interest of a few in the short term. In the longer term for our children and grandchildren, we will be leaving a planet much more depleted unstable and dangerous than the one we inherited.

As pointed out by Jackson (2009), our economic system needs serious review as it does not help us to follow sustainable development pathways, in either developed or developing countries. He explains that 'an estimated 60 per cent of the world's ecosystem services have been degraded or over-used since the mid-twentieth century'.

To put the issue in perspective, Jackson (2009) shows that if 'an estimated 9 billion people all achieve the level of affluence expected in the OECD' the global economy 'would need to be 15 times the size of today's economy (75 times what it was in 1950) by 2050' (p. 13). Based on this he questions whether this 'offers a credible vision for a shared and lasting prosperity' (p. 14).

In a similar vein, Christiana Figueres, the UNFCCC Executive Secretary, recently pointed out:

> We are on the verge of a new industrial revolution … a revolution that must succeed if we are to avoid a future of unchecked climate change. But just as the first industrial revolution depended on pumping as much carbon as possible into the air, this one that we are about to embark on will succeed only if we do the opposite (Figueres 2011).

In this book, we have tried to identify and address the challenges facing the policy makers at international and national levels to support their efforts in finding a way forward under the UNFCCC to avoid or at least to minimise these risks. Many will recognise that these challenges are not new, but we have tried to develop possible solutions to generate co-operative action with the urgency and scale required.

The important challenges are listed in the following sections and have formed the subject of each of the chapters in the book. This chapter summarises the key insights gained from the chapters and indicates some possible ways forward. There are no simple solutions and real progress involves political will and commitment to change, many stakeholders and networks working flexibly together as well as sustained effort and long-term finance.

6.2 Placing Climate Actions in a Wider Sustainable Development Context

6.2.1 The Challenge

Since the adoption of the United Nations Framework Convention on Climate Change (UNFCCC), climate negotiations have largely focused on agreement among Parties on a set of quantitative GHG emission reduction commitments. A leading principle during these negotiations has been that of *common but differentiated responsibilities*, which mean that industrialised countries should take the lead on climate change actions. This resulted in the Kyoto Protocol of 1997 with quantified commitments for industrialised countries only; developing countries were exempted from such commitments.

Attempts to extend the Kyoto Protocol with quantitative commitments for some rapidly developing countries, such as China, India, Brazil, Mexico, Indonesia and Korea, failed, and this led to a new approach in the Copenhagen Accords of 2009 and Cancun Agreements of 2010 to invite developed and developing countries to pledge GHG emission reduction actions. In Chap. 2, we could see that the pledges have been significant and diverse, but that by mid-2011, all pledges taken together would only achieve 60% of what needs to be done to keep global average temperature increases less than the 2°C target included in the Cancun Agreements (Figueres 2011).

There are several barriers to scaling up action on GHG mitigation, such as the economic crisis since 2008 and the fact that fossil fuel use is still relatively cheap (at least from a short-term perspective). Developing countries have a right and need to develop, and in that respect, climate change is seen as a constraint. Their interest is in their own sustainable development and the challenge is to design an international climate agreement in such a way that it delivers real benefits for developing countries.

This challenge has been recognised under the Convention (through the Cancun Agreements, UNFCCC 2010), and several initiatives call for action to be taken in the context of the sustainable development priorities of the country, but the modalities to implement that have not been fully established yet. The challenge is summarised as follows:

Challenge 1 To place the climate negotiations in the wider context of sustainability, equity and social change to develop agreements based on a comprehensive understanding that maximising the benefits in these areas can and should be achieved at the same time as decreasing GHG emissions.

6.2.2 Key Messages

From this challenge and the subsequent analysis in Chap. 1, the following key messages emerge:

- Action on climate change is urgent and is needed on an unprecedented scale.
- Climate and sustainable development policy making have become strongly interlinked: Climatic changes require adaptation in order that countries achieve development goals, whereas in light of the climate urgencies, there is every reason to achieve development goals with the lowest GHG emissions possible.
- There is increasing insight that for action on climate change on the scale required, a global climate coalition of all countries is needed and that that can only be achieved if countries' mitigation and adaptation actions also contribute to their sustainable development priorities and support the achievement of the Millennium Development Goals.

- Meeting a growing global energy demand with improved energy access for the poor can only go hand in hand with low GHG emission pathways. The recognition that these pathways involve rapid innovation of low-emission technologies has moved technology development and transfer to the heart of the climate negotiations and development debate.
- To assist in fulfilling the objectives of the UNFCCC and help countries to identify climate actions in the context of their sustainable development, countries can use the concepts of the updated technology needs assessment (TNA) process, low-emission development strategies (LEDS), nationally appropriate mitigation actions (NAMA) and national adaptation plans (NAP). Support can be obtained from mechanisms, such as the Technology Mechanism, the Cancun Adaptation Framework and the Capacity-Building Framework.

6.3 Towards Strategies for Climate and Development for a Green Revolution

6.3.1 The Challenges

The above challenge of placing climate actions in a wider sustainable development context led us to two further challenges:

Challenge 2 To meet both countries' sustainable development and technology transfer needs by selecting technologies or measures for climate change mitigation and adaptation based on countries' sustainable development and climate goals.
Challenge 3 To create the low GHG emission and climate-resilient strategies and action plans to accelerate innovation needed to achieve sustainable development and climate goals on the scale and timescale required.

Approaches to align climate and sustainable development actions for the identification of technology transfer needs are described in Chap. 2. How acceleration of the innovation of the portfolio of technologies or measures identified could be enabled through the creation of strategies and action plans to overcome barriers is described in Chap. 3. *These challenges are taken together here as they need to be implemented together to enable innovation on the scale required.*

It is argued that the second and third challenges could be fulfilled with the help of the provisions developed under the UNFCCC for future climate policy making, such as TNA and LEDS leading to identification of NAMAs and NAPs. To address these challenges, Chaps. 2 and 3 examine the new TNA process and compare it with what is known about LEDS, which has still to be fully specified.

6.3.2 Key Messages

This has resulted in the following key messages:

- Technologies and measures for mitigation and adaptation are key for reaching a low-emission future and for delivering sustainable development benefits provided they are implemented along with innovation support measures for overcoming barriers to their adoption.
- The new TNA process is an integrated process for both identifying and prioritising technologies and measures and for developing strategies and action plans for their rapid innovation into the country system.
- The new TNA process can be used to identify sectors that have the potential to generate sustainable development benefits and can identify a portfolio of low-emission and climate-resilient technologies or measures that can be deployed and diffused through the sector and country systems to deliver the benefits.
- The choice of technology or measure should take both climate and development benefits into account, relative to costs, rather than only climate benefits. This helps countries to maximise technologies' overall benefits compared with the costs and means that scarce resources can be allocated efficiently.
- The process is best carried out by engaging stakeholders at all levels in the country transfer system so that their knowledge is included in the analysis and their early 'buy in' in the process can be stimulated. These stakeholders can then form a network that could implement the prioritised actions.
- Based on earlier experience with TNAs conducted in developing countries, identifying technologies alone is not sufficient, but will need to be accompanied by technology innovation strategies for creating the enabling environment for the successful deployment and diffusion of prioritised technologies or measures.
- This innovation of low-emission and climate-resilient technologies or measures into a country system is a complex process requiring stakeholders to generate activities that will overcome barriers and blockages. These activities can form the basis of strategies and action plans within the country to facilitate the success of the innovation in the timescale and scale required.
- For innovations involving energy access for the poor, involvement of development non-governmental organisations (NGOs) will be needed at the early stages of transfer.
- For the implementation of technology acceleration strategies, action plans can be prepared for the short and medium to long term. These plans are costed, which will allow resource allocation for accelerating innovation as well as resource allocation for the technology or measure hardware and modification.
- A key action plan output is measuring, reporting and verification (MRV) requirements, which allows accountability and management for the actions with feedback for improvements.
- For both technology selection and strategy and action plan formulation, based on current information on the LEDS process, some harmonisation would be useful between the new TNA and the LEDS approaches to address a range of concerns.

6.4 Rationalising Current Directions in International Climate Policy Making

6.4.1 The Challenge

The Cancun Agreements contain a vision for long-term co-operative action that 'addresses mitigation, adaptation, finance, technology development and transfer, and capacity-building in a balanced, integrated and comprehensive manner to enhance and achieve the full, effective and sustained implementation of the Convention, now, up to and beyond 2012' (UNFCCC 2010, para 1).

To realise this vision, a number of concepts have been developed since the Bali Action Plan of 2007 and included in the Copenhagen Accords and Cancun Agreements, such as NAMAs (for mitigation of GHG emissions) and NAPs (supporting developing countries in taking adaptation measures). For the identification of such mitigation and adaptation actions (e.g., technologies, enabling environment improvements, training, network building, etc.), countries can formulate LEDS and/or conduct TNAs. We discussed in Chap. 3 how both TNA and LEDS could lead to a strategy for acceleration of development and transfer of prioritised technologies and measures for climate and development.

The implementation of such strategies by countries can be supported by the newly established Technology Mechanism and Adaptation Framework (UNFCCC 2011). In addition, capacity-building support could be provided through the Capacity-Building Framework under the UNFCCC, whereas through the Finance Mechanism financial support to developing countries could be facilitated.

These provisions and mechanisms are, however, still being developed. This brings us to the challenge of how they can deliver the requirements of the UNFCCC for large-scale GHG emission reduction, adaptation and capacity building in the required timescale while ensuring developing countries' sustainable development:

Challenge 4 To rationalise the current directions in international climate policy making to provide coherent and efficient support to developing countries in devising and implementing strategies and action plans for low-emission technology transfers to deliver climate and sustainable development goals.

6.4.2 Key Messages

From the analysis in Chap. 4, it became clear that:

- Outputs from the updated TNA process and LEDS can be used to produce NAMAs and NAPs. These outputs could be related to a technology level, an overarching sector or country strategy for low-emission and climate-resilient technology development and transfer with action plans.

- These outputs could form input to the Technology Mechanism and Funding Bodies for resource allocation and implementation.
- It can, therefore, be concluded that with TNA, LEDS, NAMAs and NAPs, a future climate policy regime will have important building blocks for supporting countries in formulating long-term pathways for realising environmental, social and economic development goals with low-emission and climate-resilient technologies or other measures.
- Conducting a TNA and a LEDS could support the Technology Mechanism and the Cancun Adaptation Framework and Capacity-Building Framework through this type of systematic assessment that facilitates synthesis of results across countries for regional insights useful for international policy development, MRV and resource allocation.
- Conversely, these mechanisms could support developing countries in the creation and implementation of TNA and LEDS strategies and action plans. However, to ensure efficient operation of the mechanisms in the Cancun Agreements, it is important to mitigate risk of duplication and wastage of resources. Such risk could be high, especially as the Technology Mechanism has a remit to cover both mitigation and adaptation, whereas the NAPs are the responsibility of the Cancun Adaptation Framework. Thus, options for their integration have been suggested to ensure that mitigation adaptation and capacity-building expertise is shared.
- In addition, a new step is recommended in the country process. This step would be to rationalise country strategies for technologies and measures for mitigation and adaptation and accompanying action plans to avoid overlaps and unintended duplication or interference between them before funding.
- Capacity building under the activities for acceleration of innovations would also be an output from the process, but these activities constitute part of the acceleration strategy and action plans and cannot be separated. It is, therefore, recommended that capacity building support for technology development and transfer remains within the Technology Mechanism to avoid unnecessary complexity, but again overlaps with the activities under the Capacity-Building Framework should be checked and rationalised.

6.5 Financing Technologies and Actions for Climate and Development

6.5.1 The Challenge(s)

Finally, a key prerequisite for successful climate and sustainable development pathways is availability of funding. Not only will finance needs be large, given the massive tasks that the world faces, but they will also be diverse as a consequence of country-specific needs and requirements for the purchase of equipment, improving markets or systems and poverty alleviation to ensure successful innovation. This led us to the following challenge:

Challenge 5 To facilitate development of an international framework for financial resources to support technology development and transfer, improve enabling environments for innovation, address equity issues such as poor people's energy access and make implementation of activities possible at the desired scale within the country.

In Chap. 5, we described this financial challenge that has been divided into four sub-challenges:

- The challenge to scale up finance initiatives for achieving mitigation and adaptation goals.
- The challenge to create and scale up finance initiatives for realizing development benefits, such as energy access for the poor.
- The challenge to scale up finance initiatives for purchase, modification, construction and installation of technology equipment.
- The challenge to scale up finance initiatives for accelerating development and transfer of technologies and measures for mitigation and adaptation.

In Chap. 5, we reviewed a number of studies estimating overall finance needs for mitigation in the light of the overall climate policy target of keeping global temperature increase less than 2°C and for adaptation to climatic changes. We compared these needs with funding already made available and pledged for climate actions. Experience with development aid, aid effectiveness and, in particular, finance initiatives to help poor people gaining energy access was reviewed to provide insights for financing the innovation accelerating activities and development activities. We finally discussed how funding of technologies or measures and improving their enabling environment could be acquired from public and private funding initiatives or public–private partnerships.

6.5.2 Key Messages

To scale up finance initiatives for technologies and actions for climate and development, the following key messages were found:

- Many activities have been initiated for financing mitigation and adaptation measures and sources are available from governments, multilateral and bilateral agencies, civil society, the private sector and local finance initiatives. Financing instruments used are commercial funding, subsidies, concessional loans, grants and cross-subsidisation of end-user tariffs.

- However, although climate and sustainable development are interlinked, in practice, it has turned out to be difficult for climate investment funds to simultaneously address socio-economic and mitigation and adaptation benefits: the so-called 'triple win'. Therefore, it has been difficult to find synergies in combining different public and private finance mechanisms. Improvements in financing mechanisms to ensure delivery of development benefits in addition to the climate benefits will, therefore, be needed.
- It is important that funding is not mainly focussed on projects reducing GHG emissions or enhancing adaptation, but is provided through an integrated funding strategy whereby resource needs are identified for equipment or expertise purchase, modification, operation and maintenance, as well as for creating an enabling environment for successful transfers into the country systems in the timescales and on the scale required. This observation is supported by the experience with development aid that effectiveness of financial initiatives is generally enhanced if the funding is focussed on a comprehensive strategy instead of individual projects, with active government support and commitment.
- Although funding estimates have been substantial and commitments have been made to meet these levels, most of the estimates have concentrated on low-emission technology hardware. This means that consideration will have to be given to increased funding for supporting the innovations into the country systems. Otherwise, these investments run the risk of failure and wastage of resources.
- The complexity of formulating a funding strategy in terms of appropriate sources for implementing the strategy and action plan is that the technologies or measures can have different scales of implementation and could be at different stages of development. Scale and availability in time can have implications for finance risks and this has consequences for the availability of, e.g., commercial financial resources and affordability of finance for end users.
- Private sector financial investments in low-emission and climate-resilient technologies are largely focused on technologies that are at a later stage of development (i.e., closer to commercial application). Public finance is largely used at the initial RD&D stages of technology and would also be required to support many of the innovation-accelerating activities for widespread adoption of the technologies and energy access for the poor.
- Financing technologies or measures and enabling environment improvements require an integrated funding strategy for the specific country innovation context. For this, country stakeholders can identify in the action plan what components (equipment and acceleration actions) need to be funded, and they could also suggest allocations of these funding needs across public and private finance initiatives. It is important that financial needs and resources for a technology, sector or national level strategy are rationalised to avoid duplication or gaps in funding.
- It would also be beneficial if finance support to developing countries on appropriate combinations of private and public funding instruments could come under the Technology Mechanism to provide support for a funding strategy so that implementation of action plans could be based on fully rationalised and coherent actions and funding plans. This would help to ensure consistency of

implementation over time within a country for allocations and for MRV activities. Another benefit would be the ability to aggregate the information across countries and regions for development of international supporting policies.

6.6 Moving Forward

It is not possible to devise the policies needed for a transition to a sustainable future without the political will to change the status quo. Until recently, international negotiations have so far paid little attention to whether climate actions are in accord with development interests. This has been compounded by the fact that in previous climate policy negotiations, trust has been lost. It is time to regain that trust to enable us to tackle the challenges ahead.

What we need for addressing the urgencies is both a global climate coalition and ambitious actions for enabling a rapid transition to reach our sustainable development and climate goals. The scale of the task ahead is enormous, but the risks for all of inaction are too great to ignore.

We hope that with our book, we have shown how ambitious measures can be undertaken that are fully in line with domestic interests, in both developing and developed countries, and how these measures can be supported through the international mechanisms, so that we can all address the climate change challenges as rapidly as possible.

References

Anderson, S. (2011). *Climate change and poverty reduction.* Climate & Development Knowledge Network (CDKN). Policy Brief. Retrieved August 9, 2011 from http://cdkn.org/wp-content/uploads/2011/08/CDKN_poverty-reduction_FINAL.pdf

Figueres C. (2010). *Carbon disclosure project global launch New York*, September 20, 2010. Address by Executive Secretary United Nations Framework Convention on Climate Change. Retrieved August 9, 2011 from http://unfccc.int/files/press/statements/application/pdf/100920_speech_cf_carbon_project_new_york.pdf

Figueres, C. (2011). *Statement by Christiana Figueres at the high-level inter-parliamentary debate on climate change.* Brussels, Belgium, 19 April 2011. Retrieved September 17, 2011 from http://unfccc.int/files/press/statements/application/pdf/110419_speech_inter-parliament_brussels.pdf

Jackson, T. (2009). *Prosperity without growth: Economics for a finite planet.* UK: Earthscan.

UNFCCC (2010). *The Cancun Agreements: Outcome of the work of the Ad Hoc Working Group on Long-term Cooperative Action under the Convention.* Decision 1/CP.16. FCCC/CP/2010/7/Add.1. Retrieved July 17, 2011 from http://unfccc.int/resource/docs/2010/cop16/eng/07a01.pdf#page=2

UNFCCC (2011). *Interlinkages between technology needs assessments and national and international climate policy making processes.* Background Paper III for UNFCCC workshop on technology needs assessments, Bonn, Germany, 1–2 June 2011. Retrieved June 3, 2011 from http://unfccc.int/ttclear/pdf/Workshops/Bonn%202011/Background%20Paper%20III%2025%20May%20Final.pdf

Appendix A
Lessons Learned from 'Old' TNA Process

Based on 68 completed TNAs by 2009, the UNFCCC secretariat presented a synthesis report (UNFCCC 2009). In total, these assessments had resulted in 264 project proposals based on prioritised technologies for mitigation and adaptation with potential financing needs totalling over US$ 11 billion. Most of these projects have been planned in the field of energy generation and energy saving through measures in buildings and residential dwellings. Other categories for which a relatively large number of projects have been proposed are energy efficiency measures in industrial sectors, forestry and waste management.

Project ideas in the field of climate change adaptation were mainly focused on such activities as systematic observation systems, climate change impact information and awareness building campaigns and capacity building measures. It should be noted though that the majority of TNA reports contained project concepts or ideas rather than full proposals. This was partly due to the lack of a specific format for TNA teams to prepare project proposals, but it was also because of the commercial sensitivity of publishing project data or even releasing project pipelines.[1]

In addition to project ideas and proposals, some countries included in their TNAs also programmes at the level of sectors (UNFCCC 2011). For instance, Dominica and Viet Nam identified next steps for each sector separately, while the Republic of Moldova described future strategies for the energy sector. In the field of adaptation, Turkmenistan formulated a climate observation system program. Viet Nam's sector plans contained portfolios with priority technologies, their role within the sector and preliminary estimates of the GHG emission reduction or vulnerability reduction impacts of considered technologies up to the year 2010.

---

[1] Such information may sometimes be released as part of the conditions to access certain facility or compelled due to contractual or regulatory requirements such as declaration required for statutory approval (UNFCCC 2011).

W. van der Gaast and K. Begg, *Challenges and Solutions for Climate Change*,
Green Energy and Technology, DOI: 10.1007/978-1-84996-399-2,

The plans also indicated stakeholders took part in this sector programme (UNFCCC 2011).

However, the TNA reports analysed did not show a structured approach for formulating a strategy for acceleration of prioritised technologies. Although several TNAs identified measures for overcoming barriers to technology implementation or removing system inefficiencies, these aspects were mainly addressed at the level of the projects without a clear assessment of what this would imply for an overall strategy for technology acceleration on a larger scale. Some examples of countries that identified technology implementation plans which could potentially be used as input for a national technology innovation strategy were Burundi, Democratic Republic of Congo, Madagascar and Mauritius. In the Chinese TNA, projects have been proposed as part of national or departmental development plans. According to UNDP (2008), TNAs conducted before 2008 generally gave little attention "to the non-technological options needed to create an environment conducive to the adoption of hard technologies".

Although TNAs conducted before 2008 often contained detailed overviews of priority technologies for mitigation and adaptation within prioritised sectors in the countries, with identified actions at the level of a project or sector, UNDP (2008) concluded that these actions need further development:

> to ensure more comprehensive TNAs addressing the policy aspects of technology transfer and adoption under the Convention. Linkages of TNA with key policy processes should be strengthened in future TNA analysis.

Finally, it was concluded

> that TNAs should be conducted with a clear understanding of the policy process these assessments may feed into. The information generated by TNAs, if generated through a policy-driven approach, can provide critical inputs to national efforts to address climate change (UNDP 2008, p. 8)

References

UNDP (2008). *Experiences and lessons learned from technology needs assessments (TNAs)*. Report to the Global Environment Facility (GEF). 31 May 2008. Retrieved September 22, 2011 from http://unfccc.int/ttclear/pdf/Workshops/Bonn%202011/BGPaper%20I-conducting%20TNAs.pdf

UNFCCC (2009). *Second synthesis report on technology needs identified by Parties not included in Annex I to the Convention*. FCCC/SBSTA/2009/INF.1. Retrieved June 15, 2011 from http://unfccc.int/resource/docs/2009/sbsta/eng/inf01.pdf

UNFCCC (2011). *Enhancing the implementation of the results of TNAs*. Background Paper II for UNFCCC workshop on technology needs assessments, Bonn, Germany, 1–2 June 2011. Retrieved June 3, 2011 from http://unfccc.int/ttclear/pdf/Workshops/Bonn%202011/BG%20paper%20II-implementing%20TNAs.pdf

Index

W. van der Gaast and K. Begg, *Challenges and Solutions for Climate Change*, Green Energy and Technology, DOI: 10.1007/978-1-84996-399-2,

MIX
Papier aus verantwortungsvollen Quellen
Paper from responsible sources
FSC® C105338

If you have any concerns about our products,
you can contact us on
ProductSafety@springernature.com

In case Publisher is established outside the EU,
the EU authorized representative is:
Springer Nature Customer Service Center GmbH
Europaplatz 3, 69115 Heidelberg, Germany

Printed by Libri Plureos GmbH
in Hamburg, Germany